Using Excel to Solve Statistical Problems: A Practical Guide to the Book "Statistics for Chemical and Process Engineers"

Yuri A. W. Shardt

Using Excel to Solve
Statistical Problems:
A Practical Guide
to the Book "Statistics
for Chemical
and Process Engineers"

 Springer

Yuri A. W. Shardt 🆔
Department of Automation Engineering
TU Ilmenau
Ilmenau, Germany

ISBN 978-3-031-65448-0   ISBN 978-3-031-65449-7   (eBook)
https://doi.org/10.1007/978-3-031-65449-7

This Springer imprint is published by the registered company Springer Nature Switzerland AG
The registered company address is: Gewerbestrasse 11, 6330 Cham, Switzerland

If disposing of this product, please recycle the paper.

Preface

As the usage of computers to solve complex problems becomes more commonplace, there is a need to understand how to use software to solve both basic and more complicated engineering problems, especially in statistics. One of the most common industrial software for handling data is Excel®, a programme originally developed to create spreadsheets and handle basic accounting operations that has since expanded to include more complex capabilities. In the industrial world, Excel is often the default engineering standard for data processing and manipulation. Thus, there is a need to understand how to implement the various statistical concepts easily and effortlessly in Excel.

This companion book is meant to focus on providing the examples and details for solving statistical examples using Excel. The theoretical details and background information can be found in the author's acclaimed book *Statistics for Chemical and Process Engineers: A Modern Approach* (https://link.springer.com/book/10.1007/978-3-030-83190-5) now in its second edition. The companion book is structured so that the individual chapters reflect the material presented in the original book, which makes looking for the details easier and faster.

In order to show clearly the Excel code and functions, these will be typeset in `Courier New`. It assumed that all cells have been labelled using the variable provided in square brackets before the equals sign. As well, it is assumed that in tables and examples, the Excel variables flow from top to bottom which means that a variable previously defined in a row or line of code above will mean the same later on, for example, if we see `[A] = sqrt(5)`, then later when we see `[B] = A^2`, we can assume that the reference is to the same A. In tables, the user input is defined at the bottom of the table and will normally refer to the variables as defined by some mathematical equations. Finally, if a cell is defined as a specific location, for example, C1, then it can be assumed that when being dragged down or across cells, the values will change based on the definition in the formulae.

Erfurt, Germany   Yuri A. W. Shardt

Contents

List of Figures

List of Tables

Introduction to Excel® 1

This section presents a brief overview of Excel and its key features. It assumes that the reader is sufficiently familiar with Excel to open files, enter values into a cell, and navigate between the different components. Advanced knowledge of Excel is not required. Additional information can be found in such resources as *Excel for Dummies* (Harvey 2021).

The basic Excel interface is shown in Fig. 1.1. The main features that will be used in subsequent chapters are highlighted and briefly explained below:

A. **Excel Ribbon**: The Excel ribbon contains a series of tabs that contain useful functions for different tasks. We will be looking at specific functions and ribbons in each chapter.
B. **Cell/Range Name**: This little window is extremely useful in that it allows the user to define a specific name to a given cell or range.
C. **Formula Window**: This is where all the formulae are entered in Excel. A formula starts with an equals sign (=) in Excel.

A **cell** is a single block in Excel that contains numbers, text, or formulae. A cell is referred to by a column letter and a row number, for example, cell B2 is the cell in column B and row 2. A cell is **referenced** (referred to) by its cell name. Referencing can be either **relative** or **absolute**. Relative referencing means that the cell location changes as a formula is dragged to different cells. Absolute referencing means that the same cell is always referred to. Absolute referencing is denoted by placing a dollar sign $ before the component that is not to change. There are three possibilities:

© The Author(s), under exclusive license to Springer Nature Switzerland AG 2024   1
Y. A.W. Shardt, *Using Excel to Solve Statistical Problems: A Practical Guide to the Book "Statistics for Chemical and Process Engineers"*,
https://doi.org/10.1007/978-3-031-65449-7_1

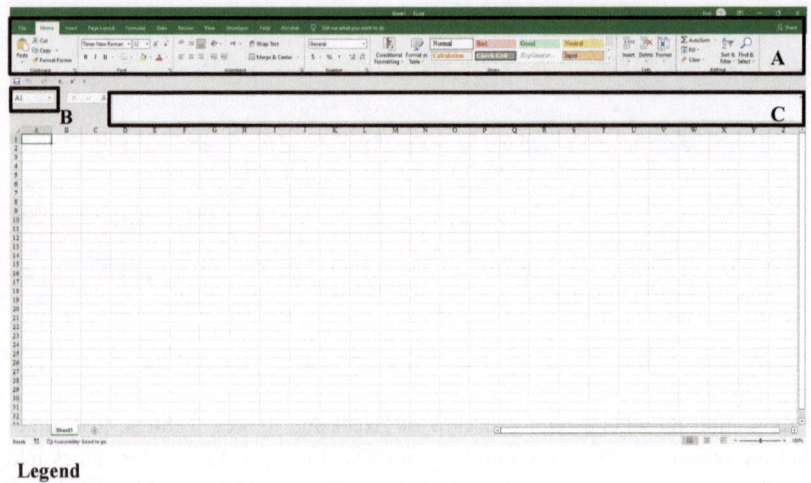

Legend

A: Excel Ribbon   B: Range/Cell Name   C: Formula Window

Fig. 1.1   Excel interface (Excel 2019)

(1) Completely absolute referencing with a $ before each component, for example, B2.
(2) Absolute column with a $ placed only before the column component, for example, $B2. This fixes the column, but allows the row to vary.
(3) Absolute row with a $ placed only before the row component, for example, B$2. This fixes the row, but allows the column to vary.

A **range** is a rectangular group of cells that are to be analysed or treated together. A range is specified by the start and end cells using a colon in between, for example, B2:C4 represents a range starting with cell B2 and going to cell C4 with a size of 2 columns (B and C) and three rows (2, 3, and 4). Nonadjacent cells can be combined using a comma, for example, the range denoted by B2: C4, F1:F5 consists of two nonadjacent parts. A range can also be selected using the Ctrl button. Select the first range using the mouse. Press and continue holding the Ctrl button. Select the next range. When finished release the Ctrl button.

Arrays are a range of adjacent cells that behave in Excel similarly to matrices. There are special Excel functions that allow us to manipulate arrays as if they were matrices and obtain matrix results.

Table 1.1 Summary of the basic Excel mathematical commands with examples (Let [A1] = 1 and [A2] = 4.)

Command	Definition	Example	Output
+, −, *, /	Add, subtract, multiply, divide	= A1 + A2	5
^	Exponentiation	= A2^2	16
sqrt()	Square root	= sqrt(A2)	2

Cells, ranges, and arrays can be named for easier access in the future. Select a cell, array, or range and go to the window marked by B in Fig. 1.1 and enter a name for the selected range. This name can then be used instead of the range reference. As well, this approach has the advantage that when adding columns and rows between the endpoints of the range, Excel will automatically update the range referred to by the specific name. User-defined names should be unique to each spreadsheet to avoid unexpected behaviour.

1.1 Basic Excel Commands

Table 1.1 presents some commonly used mathematical operators in Excel with brief examples.

1.2 Array Functions to Create Matrices and Vectors

Arrays are defined as a range of cells that are treated together. In effect, this allows Excel to handle matrices and vectors. **Array functions** allow the arrays to be manipulated as if they were matrices or vectors. However, the user must self-define the output array that will contain the solution. When using array functions, the following procedure must be followed:

(1) Select the output range.
(2) Enter the array formula into one of the cells in the selected array.
(3) Once the formula has been entered press Ctrl + Shift + Enter to register the formula as an array formula. Normally, one would simply press Enter.
(4) Excel will place curly brackets around the array formula. Note that one cannot edit part of an array formula.

Table 1.2 Excel array functions

Function	Description
mdeterm(array)	Determines the determinant of an array. The result will be a single scalar value
minverse(array)	Determines the inverse of the $n \times n$ array. The result will be the same size as the initial array
mmult(array1, array2)	Multiples two arrays array1 and array2 together. If array1 has size $m \times n$, then array2 must have size $n \times p$. The result will have size $m \times p$
transpose(array)	Transposes an array, that is, the rows and column are exchanged. If the array was originally $n \times m$, then the output will be $m \times n$

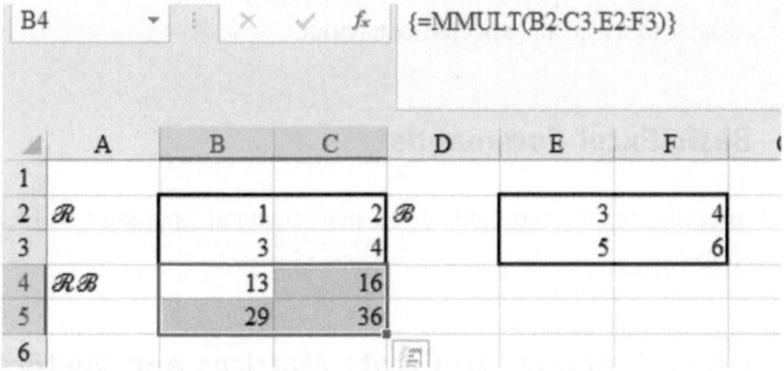

Fig. 1.2 Using array functions in excel

A summary of the most common array functions is given in Table 1.2. An example of how to use array functions is shown in Fig. 1.2. The two matrices are defined and then multiplied together using the appropriate array function. The array must be selected before the function is entered.

1.3 Excel Macros and Security

Macros are Excel's version of functions, or user-written code, that Excel can execute. The programming language used by Excel is called Visual Basic (VBA).

Table 1.3 Logic operators in Excel VBA

Logic statement	Operator	Logic statement	Operator
Equals	=	Less than	<
OR	Or	Less or equal to	<=
AND	And	Greater than	>
NOT	Not	Greater than or equal to	>=
Not equal to	< >		

In Excel 2007 or newer, code can be inserted by going to the View Ribbon, selecting the Macro icon, and then View Macro. In the window that appears, enter the name of the function that you desire to create (or edit) and press Create (Edit). If a new function is being created then, in the new window that opens, replace Sub with Public Function. This will allow the new code to be directly accessed from the spreadsheet by typing = FunctionName(Required Parameters). Below, some sample code has been provided that implements the Michaelis–Menten equation.

```
Public Function MichaelisMenten(Concentration, vmax, KM)
As Double
'This function will contain a single line of code that
implements the Michaelis-Menten equation
MichaelisMenten = vmax * Concentration / (KM + Concentra-
tion)
End Function
```

Common Excel logic operators are given in Table 1.3, while Table 1.4 gives the common Excel VBA branching commands.

1.3.1 Security in Excel

Unfortunately, when a macro is designed, Excel has the tendency to be paranoid and think that it is always a nasty virus. Thus, the appropriate parameters should be set for security. The procedure in dealing with security in Excel depends on

Table 1.4   Branching commands in Excel VBA

Command	Comments

`if (logic) Then`
`(statement)`
`Else`
`(statement)`
`End If`
	If–then statement. The `Else` can be replaced by `ElseIf…Then` to provide a chain of possibilities

`For I = m To p Step n`
`(statement)`
`End;`
	For-loop, the increment n can be skipped, in which case the default increment of $+1$ is used

`While (logic)`
`(statement)`
`End;`
	While-loop

`Return`	Allows one to exit a loop prematurely

the version of Excel installed. The following sections explain the procedures for Excel 2016 (and newer).

In Excel 2016 or newer, to set the security, go to the `File` menu, and select `Options`. Select `Trust Center` in the window that appears. After this, select `Trust Center Settings…`. In the new window, go to `Macro Settings` and select the appropriate level of security you desire. A good choice is to select the option `Disable All Macros with Notification` because the macros will be disabled, but you will be notified of their existence. Press `OK` on all the open Windows to save the changes. A file with a macro must be saved as an `.xlsm` file. When opening the file with a macro and the above-suggested settings, a warning similar to that shown in Fig. 1.3 should appear. Clicking on `Enable Content` will permit the macros to be used.

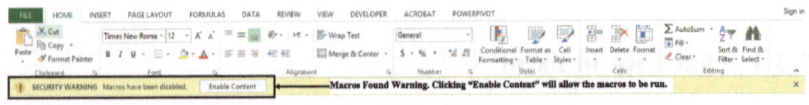

Fig. 1.3   Security warning when macros are present (Excel 2019)

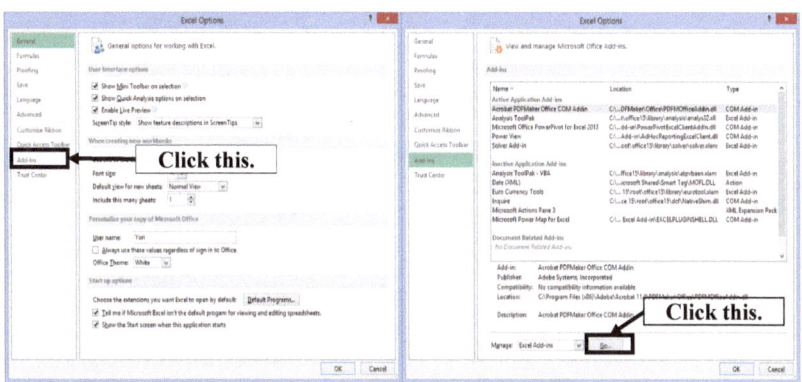

Fig. 1.4 Navigating to the solver installation menu (Excel 2019)

1.4 Excel Add-Ins

There are two very useful Excel add-ins, the Solver and the Data Analysis Add-Ins, that can make using Excel easier. Unfortunately, they need to be installed specially in order to be accessible. Details on how to install and use them are provided in the sections below.

1.4.1 The Excel Solver Add-In

Solver is an Excel add-in that allows the user to iteratively solve systems of equations. Unfortunately, it is not installed by default on most computers.

1.4.1.1 Installing the Solver Add-In

In Excel 2016 or newer, in order to install the Solver add-in, go to the File menu and select Options at the bottom of the Menu that appears. In the new window that appears, select Add-ins. Finally, click the Go... button. The last two steps are shown in Fig. 1.4. A window similar to Fig. 1.5 should appear.

1.4.1.2 Using the Solver Add-In

In order to start Solver, in Excel 2016 or newer, locate the Data ribbon and go to the extreme right-hand side in the area marked Analysis. Solver should be there as shown in Fig. 1.6.

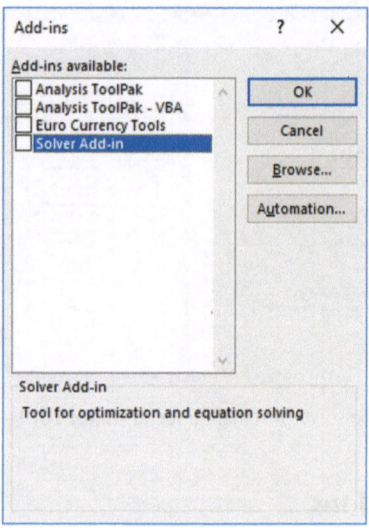

Fig. 1.5 Installing solver

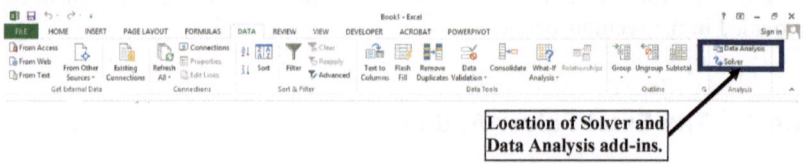

Fig. 1.6 Location of the Solver and Data Analysis add-ins (Excel 2019)

Figure 1.7 shows the Main Solver Window that appears in Excel 2016 or newer. It is a must that the option `Make Unconstrained Variables Non-Negative` be **unchecked**, as it can lead to wrong results otherwise. The following sections are important for use in regression analysis:

(1) **Objective Function Value**: This is the value of the objective function that is to be optimised.
(2) **Type of Optimisation**: What type of optimisation is desired: maximisation (`Max`), minimisation (`Min`), or force the solver to obtain a particular value (`Value of`). For regression, the minimisation option should be used.

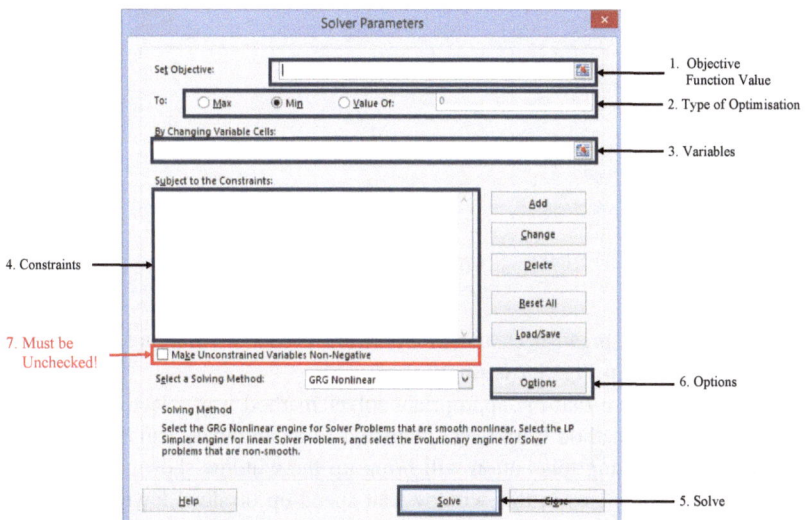

Fig. 1.7   Main Solver Window (Excel 2016 or newer)

(3) **Variables**: This is the range of the cells (variables) that the computer can vary to determine the solution. For regression, this would represent the cells where the parameter values have been entered.

(4) **Constraints**: This box lists the constraints for the problem. In order to add a constraint, click on the Add button. The window shown in Fig. 1.8 should appear. Once the desired form of the constraint has been selected, click Add to add the constraint to the list of constraints. Selecting a constraint from the box and clicking Change will cause the same window to appear and the properties of the constraint can be changed. Finally, selecting a constraint and clicking Delete will remove the constraint.

(5) **Solve**: Clicking this button will start the solver. The solution of the problem may take some time. Solver will either state that a solution was found (Fig. 1.9 (left)) or that no solution was found (Fig. 1.9 (right)). In general, if a solution is found, select Keep Solver Solution and press OK; otherwise select Restore Original Values and press OK. If the Solver fails to find a solution, an error message will be included. It can give a suggestion as to how to fix the problem. Three common things to check (in order of preference) are that:

   a. The number of iterations was not exceeded;

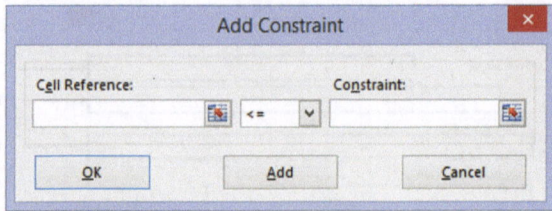

Fig. 1.8 Add constraint window

    b. The Excel spreadsheet and Solver were properly configured, especially that Box 7 in Fig. 1.7 is **unchecked**; and

    c. To make sure that the appropriate solver method was selected. Changing the solver method from GRG nonlinear to evolutionary can be useful.

(6) **Options**: Clicking this button will bring up the window shown in Fig. 1.10. Each of the choices in this window can speed up or slow down the amount of time required to obtain a solution or even if a solution can be found. Each option will be discussed separately:

    a. **Max Time**: This represents the maximum amount of time that Solver will run in order to determine a solution. If the problem is large, then increasing this option can potentially allow Solver to find a solution.

    b. **Iterations**: This represents the maximum number of iterations that Solver will perform before it stops. If the initial guess is far from the solution, it may take many iterations before a solution is obtained. Thus, increasing the number of iterations can be a good idea.

    c. **Precision**: This represents the largest possible difference between the calculated value of the constraints and the specified value of the constraints. The smaller the number the longer it will take to find a solution.

    d. **Tolerance**: This is similar to precision but is used for integer constraints. It represents the percentage by which the calculated values differ from the specified values.

    e. **Convergence**: This is similar to precision but is used to compute the maximum allowable difference between two iterations of the parameters (or cells that can change). Since for most purposes, a relative value would be better, this entry should be changed whenever the parameters are expected to either be all very large numbers or very small numbers.

    f. **Use Automatic Scaling**: This should always be selected as it minimises the effect the magnitude of the different variables can have on the solution.

It is especially important if one of the variables ranges from 100 to 1,000, but the other variable ranges from 0.01 to 1.

The options in the other tabs are mostly irrelevant and should be left at their default values unless the problem at hand requires special treatment. However, the correct approach to take requires consulting an appropriate source on numerical methods.

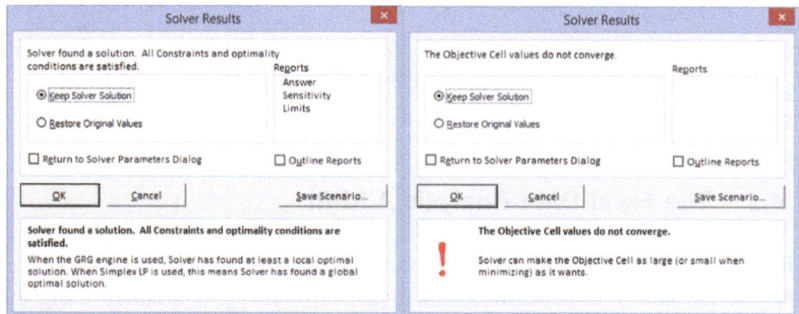

Fig. 1.9 (Left) Solver found a solution and (right) Solver failed to find a solution (one possible result)

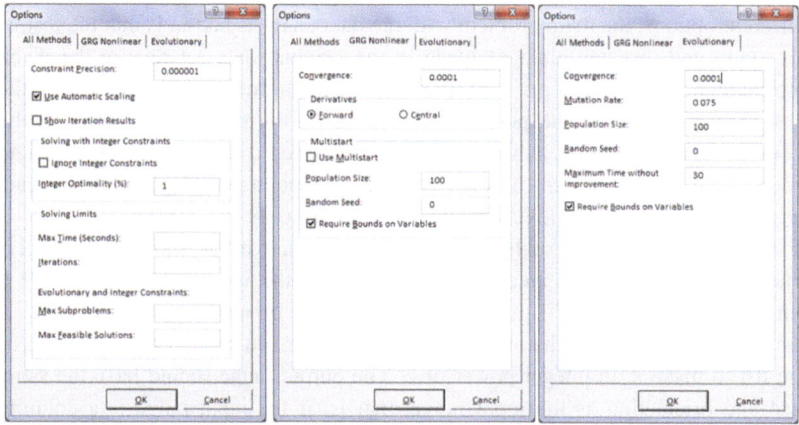

Fig. 1.10 Solver option window (Excel 2016 or newer)

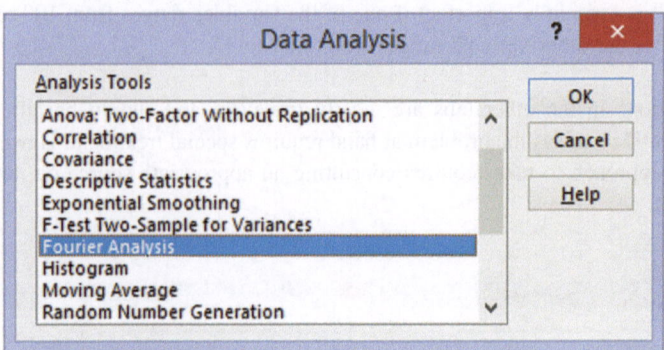

Fig. 1.11 Data analysis window (Excel 2016 or newer)

1.4.2 The Excel Data Analysis Add-In

The Data Analysis add-in in Excel is another very useful Excel add-in that can improve the ability to perform certain statistical tasks. It is installed using the same procedure as installing the Solver add-in (see Sect. 1.4.1.1: Installing the Solver Add-In). In order to start the Data Analysis Add-In, locate the Data ribbon and go to the extreme right-hand side in the area marked Analysis. The Data Analysis Add-In should be there as shown in Fig. 1.6.

The Data Analysis window is shown in Fig. 1.11. Although there are many different options, the main problem with the data analysis add-in is that the results are static and that any changes made in the original data set require the given programme to be rerun. As well, the display of information is not always the best. Nevertheless, for the purposes of this book, the only useful option is the **Fourier Analysis** option, which will compute, given a data set, the appropriate Fourier coefficients, which can then be used to create a periodogram for the data set. An Excel template file has been created to simplify the process.

Selecting the Fourier Analysis option will give the window shown in Fig. 1.12. There are only two key areas to consider. First, the input range must have a length of 2^n, where $n \in \mathbb{N}$, that is, the length must be an integer power of 2. If the particular list is less than the desired value, then add extra zeros to the end of the list to make it an integer power of 2. The output range should have the same size and orientation as the input range, that is, if the input range is a column, then the output range should also be one; similarly for a row. Clicking OK will give the required Fourier coefficients.

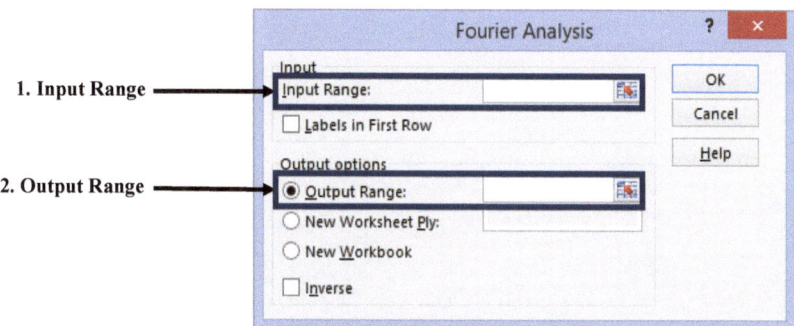

Fig. 1.12 Fourier analysis window (Excel 2016 or newer)

Reference

Harvey, G. (2021). *Excel® All-in-One For Dummies*. Hoboken, New Jersey, United States of America: John Wiley & Sons, Inc.

Reference

Data Visualisation

2

This chapter presents material associated with Chap. 1: Introduction to Statistics and Data Visualisation *of the book* Statistics for Chemical and Process Engineers: A Modern Approach.

2.1 Summary of Relevant Functions

2.1.1 Excel Functions for Descriptive Statistics

Given an array x that contains the data that is to be analysed, the functions in Table 2.1 can be used to compute the various commonly used descriptive statistics.

2.1.2 Data Visualisation in Excel

Excel contains many prebuilt procedures that can be used to create graphs. In Excel, graphs are called **charts** and can be found in the Insert ribbon under the Charts section. There are many different types of possible charts, including bar (both vertical and horizontal), pie, scatter, line, statistical, and various combinations of these. Figure 2.1 shows the Insert ribbon and where the different charts can be found. An in-detail look at the Charts section is shown in Fig. 2.2. The key types of graphs have been labelled by letters: A for bar graphs; B for line graphs; C for statistical graphs including histograms and box-and-whisker plots; and D for various types of scatter graphs.

© The Author(s), under exclusive license to Springer Nature Switzerland AG 2024   15
Y. A.W. Shardt, *Using Excel to Solve Statistical Problems: A Practical Guide to the Book "Statistics for Chemical and Process Engineers"*,
https://doi.org/10.1007/978-3-031-65449-7_2

Table 2.1 Useful Excel functions for descriptive statistics (Let x be an array of 10 random numbers generated using the `RAND()` function in Excel. Note that the values obtained will differ depending on the run. There is no way to force a particular set of numbers in Excel)

Command	Definition	Example	Output
`average(x)`	Mean value of x	`=average(x)`	0.441344
`median(x)`	Median value of x	`=median(x)`	0.46011
`stdev.s(x)`	Standard deviation of x	`=stdev.s(x)`	0.322008
`var(x)`	Variance of x	`=var(x)` `=stdev.s(x)^2`	0.103689 0.103689
`skew(x)`	Skewness of x	`=skew(x)`	0.347016
`max(x)-min(x)`	Range of x	`=max(x)-min(x)`	0.992928
`quartile.exc(x,k)`	Computes the kth quartile for x	`=quartile.exc(x,1)`	0.143328

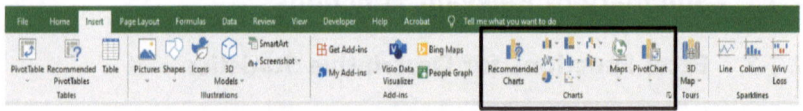

Fig. 2.1   Insert ribbon in Excel

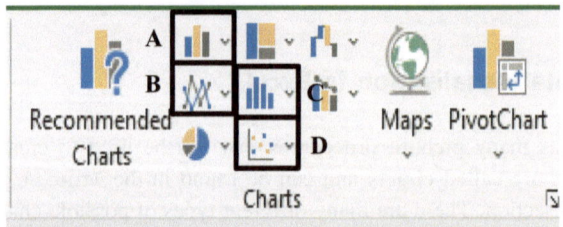

Fig. 2.2   Chart section in Excel highlighting the key types of graphs

In all cases, the data should be first entered into Excel and then select the desired graph. Excel will then allow you to customise the graph and make it look more scientific. It is important to remember to place labels for axis and legends.

Although Excel can create many different types of graphs, there still are certain types that must be created by the user. For example, normal probability plots cannot be created from scratch by Excel. Instead, the user must create the individual components separately and plot the final graph. Excel templates for normal probability plots and periodograms are available.

2.1.2.1 Normal Probability Plot Template

Requirements: Basic Excel installation

Goal: Create a normal probability plot in Excel. Can be modified to deal with other distributions.

File Name: `normplot.xltx`.

Description: A screenshot of the template with an explanation of the formulae used is shown in Fig. 2.3. The resulting normal probability plot is shown in Fig. 2.4. The steps for creating a normal probability plot can be summarised as follows:

(1) Place the original data in Column A.
(2) Obtain the order of the data in Column A in Column B. You can use the rank function.
(3) In   Column   C,   enter   =   `norm.s.inv((ColumnB1-0.5)/count(Column$A))`. The value of 0.5 is subtracted from the original ranked value in order to avoid asking the computer for the location for which the probability is 100% (it is $+\infty$!).
(4) In Column D, compute the Z-score for each of the data points, that is, subtract the mean and divide the resulting value by the standard deviation of the values in Column A.
(5) Plot a scatter plot of the data in Columns C and D.
(6) The straight line can be added by plotting the data in Column C against itself.

Warnings: The axes of the plot are fixed to the range $[-3.0, 3.0]$. Should there be data outside this region, then it will be necessary to manually change the axis limits.

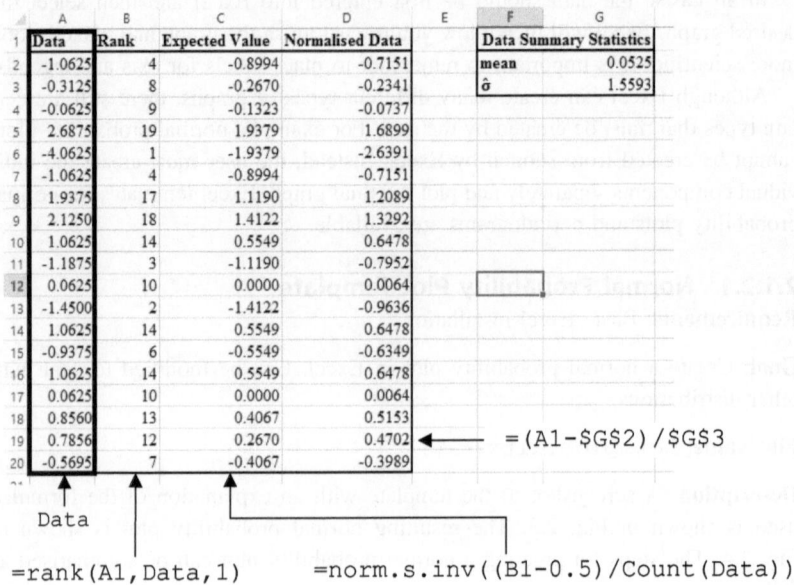

Fig. 2.3 Normal probability plot data (The formulae given are those placed in the first row, they would then be dragged down into each of the remaining rows.)

2.1.2.2　Periodogram Template

Requirements: Basic Excel installation plus installing the Data Analysis add-in (see Sect. 1.4.2: The Excel Data Analysis for how to install it.)

Goal: Create both the full and half periodograms in Excel.

File Name: `periodogram.xltx`.

Description: A screenshot of the template is shown in Fig. 2.5 with the resulting periodograms shown in Fig. 2.6. Note that every time new data are entered, it is necessary to rerun the Fourier analysis function in the Data Analysis add-in. The set-up of the Fourier transform window is shown as an inset in Fig. 2.5. As well, the number of data points must be a multiple of 2^n where n is an integer, that is 2, 4, 8, 16, 64, 128, 456,.... If the data set of interest is not a multiple, then it is necessary to add extra zeros to the end of the list to make it so.

An explanation of the columns is as follows:

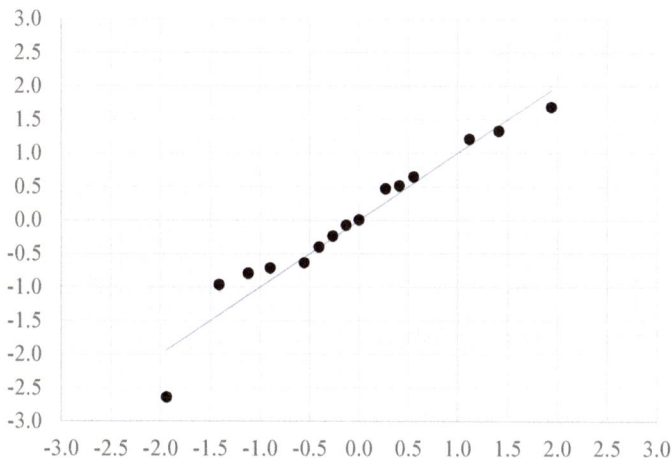

Fig. 2.4 Resulting normal probability plot

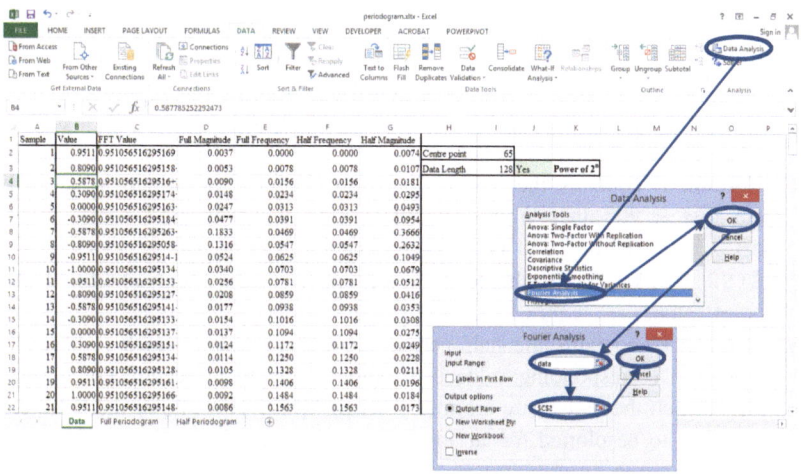

Fig. 2.5 Periodogram template layout (Excel 2019). The inset shows how to initialise the Fourier analysis function

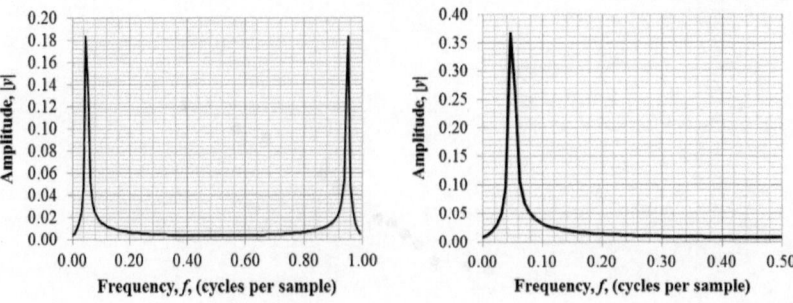

Fig. 2.6 Sample full and half periodograms

(1) **Column A** contains a simple count of the sample number **starting from 1**.
(2) **Column B** contains the values corresponding to each sample number. This column is called **data** and must be a multiple of 2^n, where n is an integer.
(3) **Column C** contains the Fourier transform values as returned by the Fourier analysis function in Excel. The values are complex numbers and should not be changed.
(4) **Column D** contains the magnitude of the values in Column C, that is, $=$ abs (C2). This column is used to construct the full periodogram.
(5) **Column E** contains the frequency corresponding to each sample, that is, $=$ (A2-1)/COUNT(data).
(6) **Column F** contains the half periodogram frequencies, which is basically the first 2^{n-1} values from Column E with the remaining values set to #N/A, so that they will be ignored. The formula used is $=$ IF(A2-1 < I2,E2,#N/A). It should be noted that cell I2 contains the centre point value.
(7) **Column G** contains the half periodogram magnitudes, which is basically twice the corresponding value in Column D, up to the centre point value, after which the values are arbitrarily set to #N/A. This allows the half periodogram to be plotted for an arbitrary number of values. The formula used is $=$ IF(A2-1 < I2,D2*2,"NaN").
(8) **Full Periodogram**: The full periodogram is created by plotting Column D as the y-axis and Column E as the x-axis.
(9) **Half Periodogram**: The half periodogram is created by plotting Column G as the y-axis and Column F as the x-axis.

Warnings: The Fourier transform function must be rerun each time the data are changed. Furthermore, the data length must always be a multiple of 2^n, where n is an integer.

2.1.3 Excel Chart Ribbon

When a chart is selected in Excel, a Chart Tools superribbon appears that consists of two ribbons: Chart Design and Format, which are shown in Fig. 2.7. From our perspective, there are few useful features that we will need to use. These have been highlighted in Fig. 2.7. Most of the elements are self explanatory. Move Chart, which is marked as G, allows the location of the chart to be changed by being moved to a different spreadsheet or as its own spreadsheet. In general, when copying charts from Excel into Word or another programme, it is better that they be located on their own spreadsheets.

Additional windows appear when selecting specific elements in the chart. These will be explained in the detailed example.

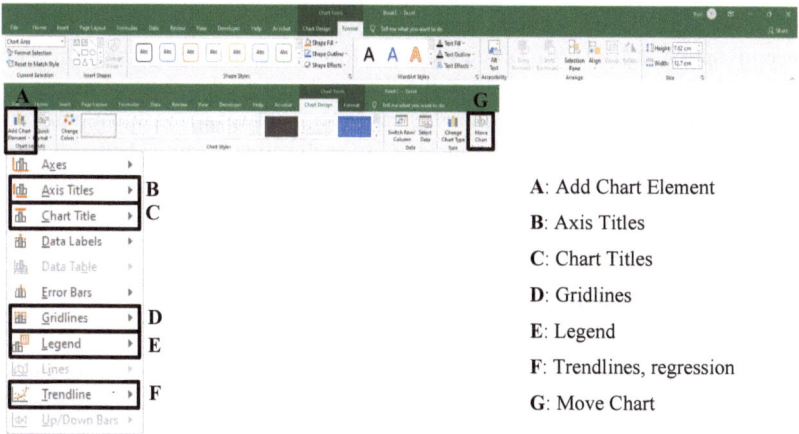

A: Add Chart Element

B: Axis Titles

C: Chart Titles

D: Gridlines

E: Legend

F: Trendlines, regression

G: Move Chart

Fig. 2.7 Chart tools superribbon in Excel

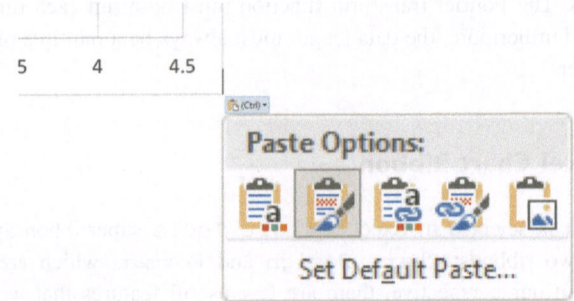

Fig. 2.8 Pasting Excel charts in word

2.1.4   Copying Charts

Excel allows charts to be easily copied from one Microsoft programme into another. When copying charts, it is possible that the formatting will be updated to match that of the destination document. If this feature is undesired, then it should be removed by right-clicking on the paste-options drop-down menu that appears in the bottom right-hand corner when hovering over the figure and selecting the second option `Keep Source Formatting & Embed Worksheet`. This will allow you to make changes in the data and update the chart as necessary. However, certain formatting features (such as decimal separators) may not convert properly between different computers. In such cases, it can be better to paste the chart as a fixed figure in Word (or Powerpoint). This requires selecting the last option `Picture` (Fig. 2.8).

2.2   Detailed Example

Using the data in Table 2.2 that shows the relationship between the measured and theoretical freezing-point values of compound A at different concentrations, create a graph that displays the freezing point as a function of the concentration. Make sure to label all axis and provide a legend if necessary.

Table 2.2 Theoretical and experimental freezing-point values at different concentrations of compound A

Concentration A	Freezing point (°C) theoretical	Freezing point (°C) experimental
0.00	−60.00	−58.61
0.05	−58.00	−63.01
0.10	−54.00	−56.90
0.15	−53.00	−42.70
0.20	−45.00	−38.41
0.25	−40.00	−38.37
0.30	−36.00	−33.57
0.35	−32.00	−32.04
0.40	−24.00	−18.38
0.45	−15.00	−12.59
0.50	−10.00	−9.92
0.55	−7.00	−6.52
0.60	−6.00	−5.72
0.65	−5.00	−3.02
0.70	−4.00	−6.79
0.75	−3.00	−0.06
0.80	−2.00	1.66
0.85	−1.00	−0.70
0.90	0.00	1.68
0.95	1.00	2.83
1.00	2.00	2.72

2.2.1 Prerequisites

Copy the above data into an Excel spreadsheet. For our purposes, we will assume that the data is found in the range A1:C22, as shown in Fig. 2.9.

Note that Excel always assumes that the left-hand column contains the x-values. Therefore, it helps to have the first column in the range as the x-values.

	A	B	C
1	Concentration A	Freezing Point (°C) Theoretical	Freezing Point (°C) Experimental
2	0.00	-60.00	-58.61
3	0.05	-58.00	-63.01
4	0.10	-54.00	-56.90
5	0.15	-53.00	-42.70
6	0.20	-45.00	-38.41
7	0.25	-40.00	-38.37
8	0.30	-36.00	-33.57
9	0.35	-32.00	-32.04
10	0.40	-24.00	-18.38
11	0.45	-15.00	-12.59
12	0.50	-10.00	-9.92
13	0.55	-7.00	-6.52
14	0.60	-6.00	-5.72
15	0.65	-5.00	-3.02
16	0.70	-4.00	-6.79
17	0.75	-3.00	-0.06
18	0.80	-2.00	1.66
19	0.85	-1.00	-0.70
20	0.90	0.00	1.68
21	0.95	1.00	2.83
22	1.00	2.00	2.72
23			

Fig. 2.9 Entering the data into Excel

2.2.2 Solution

Select the data in cells A2:C22. We will use the headers later. Go to the Insert ribbon and select the Scatter Plot icon (D in Fig. 2.2). Click on the drop-down menu and select the first option. This will create the default scatter plot in the current worksheet (which we can later move if desired). The resulting graph should look like that shown in Fig. 2.10.

Now, it is necessary to properly format the two data series so that we have the experimental values as black dots and the theoretical values as a solid line. We will start by selecting the blue dots and right-clicking on them. In the drop-down menu that appears, select Format Data Series. This is shown in Fig. 2.11. A format data series window will appear on the right. In this window, select

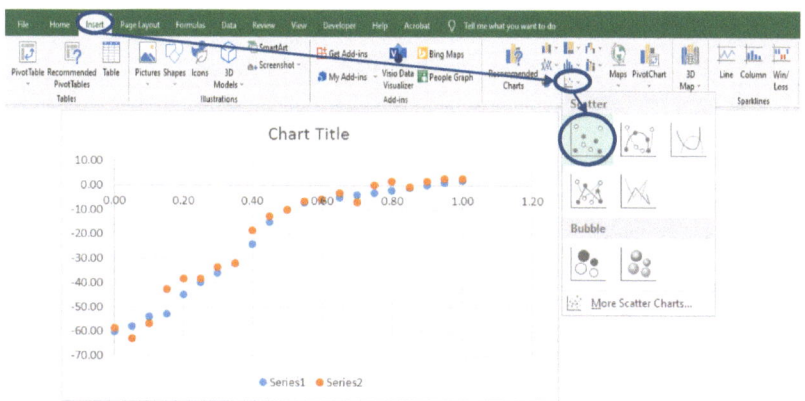

Fig. 2.10 Creating a graph in Excel

the Fill and Line tab. In the tab, select the Line subtab. This brings up attributes associated with the line property of the data series. Since we want a solid line, select the Solid line option. Set the colour to black and the line width to 1 pt. The graph updates as you make the changes. These steps are shown in Fig. 2.12 (left). Now, we need to remove the marker (dots) for the theoretical data series. This can be accomplished from the same Format Data Series window by clicking on the Marker subtab. Click on the Marker Options section and select None. This is shown in Fig. 2.12 (right). This will give a chart that looks like that shown in Fig. 2.13.

Now, we need to modify the experimental data (orange dots). We will follow a similar procedure as for the theoretical values but certain changes will not need to be made. Select the orange dots and right-click on one of the orange dots. Select Format Data Series and in the window that appears select the Fill and Line tab. Finally, select the Marker tab. On the Marker tab, select the Marker options section and select the Built-in radio button. This option allows us to directly influence how the given dot appears. We can change the type of marker by changing the selection in the Type drop-down menu. Here we will not make any changes to the type. However, we will change the size by clicking on the Size box and entering 3. Next, we will select the Fill section and change the colour. First, we select Solid fill and then set the colour to black in the Color menu. Finally, we will set the border of the marker using the Border section. Select the Solid line radio button and set the colour to

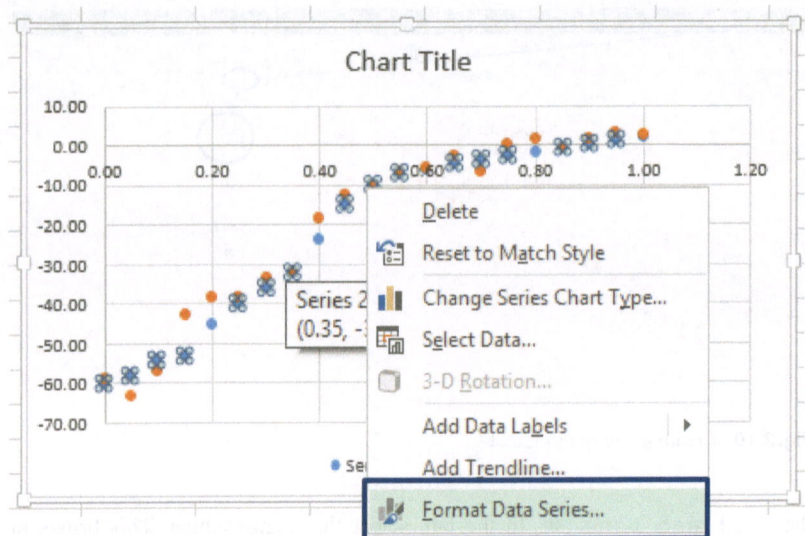

Fig. 2.11 Selecting a data series and formatting it

black in the colour menu. Finally, set the width to 0.75 pt in the Width menu. You may need to scroll down in order to see all the options. This is shown in Fig. 2.14. The resulting graph is shown in Fig. 2.15.

Once the data has been properly formatted, it is necessary to make the legend meaningful. This can be accomplished as follows. Select and right click on a black data point. Select the Select Data... item. The Select Data Source window will appear. Select Series 1. Click the Edit button. In the window that appears in the Series Name field, enter Theoretical Value. Press OK. This is shown in Fig. 2.16. Select Series 2 and repeat the previous step but enter the name Experimental Values. The series name can be a cell reference. The checkmark (✓) beside a series name shows that a given data series will be plotted.

Next, the axes need to be properly formatted. This includes adding labels, placing them appropriately, and formatting as necessary. For the x-axis, right click on the x-axis and select Format Axis.... The Format Axis window will appear on the right-hand side of the window. Select the Axis Option tab (the one with a bar graph). Change the maximum bound to 1.0. Excel tries to determine reasonable bounds automatically. However, it may not provide the best

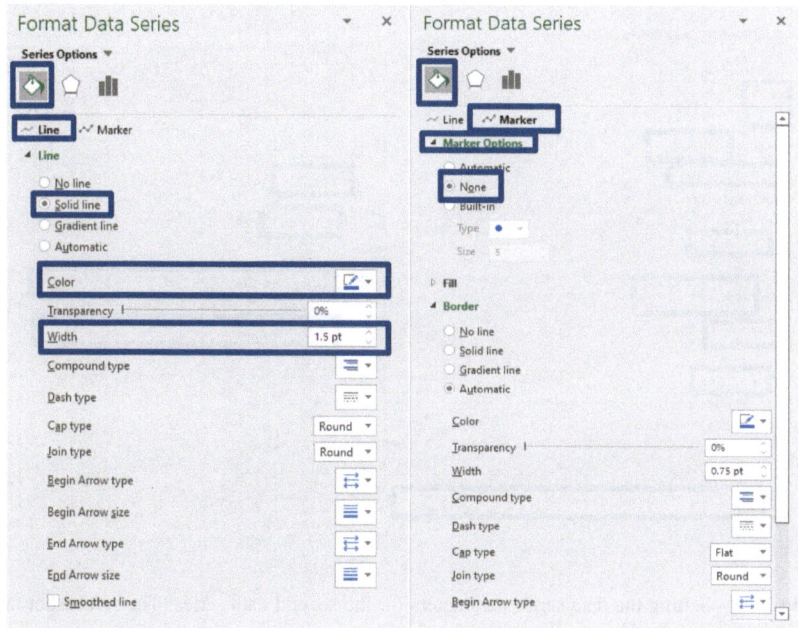

Fig. 2.12 (Left) Defining the line properties and (right) Defining the marker properties for the data series

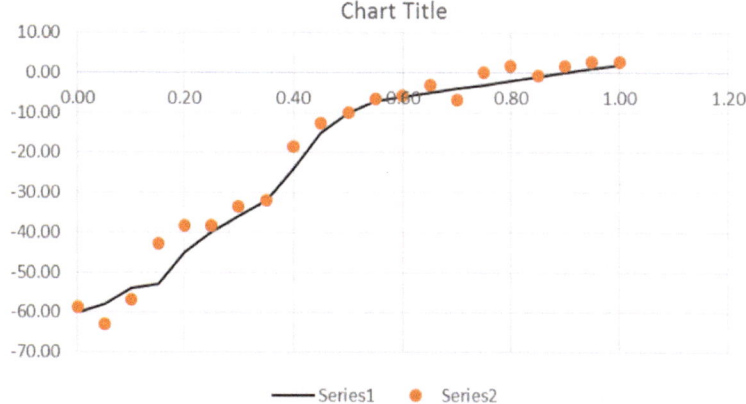

Fig. 2.13 Chart after modifying the theoretical values

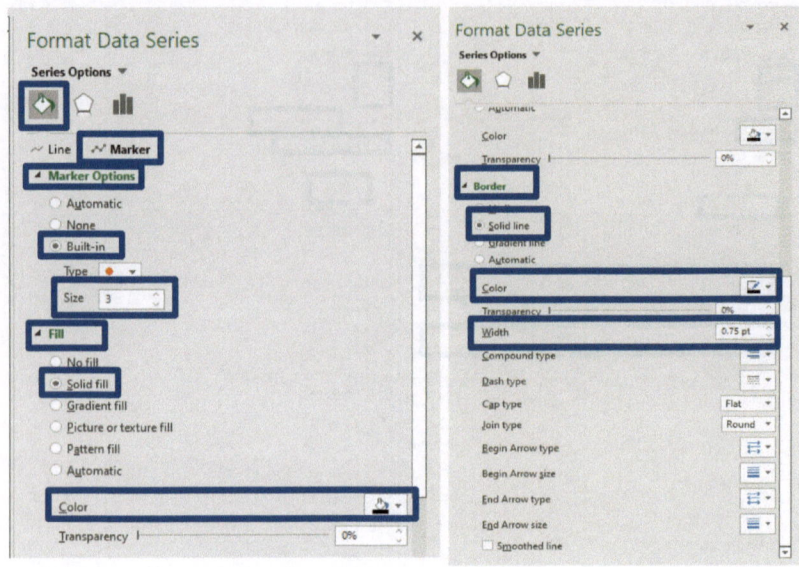

Fig. 2.14 Setting the data series parameters for the second data series (The screenshot on the right is obtained by scrolling down.)

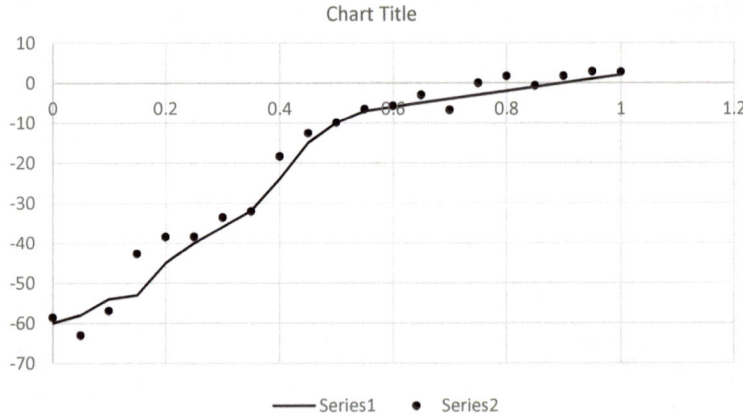

Fig. 2.15 Chart after formatting the experimental values

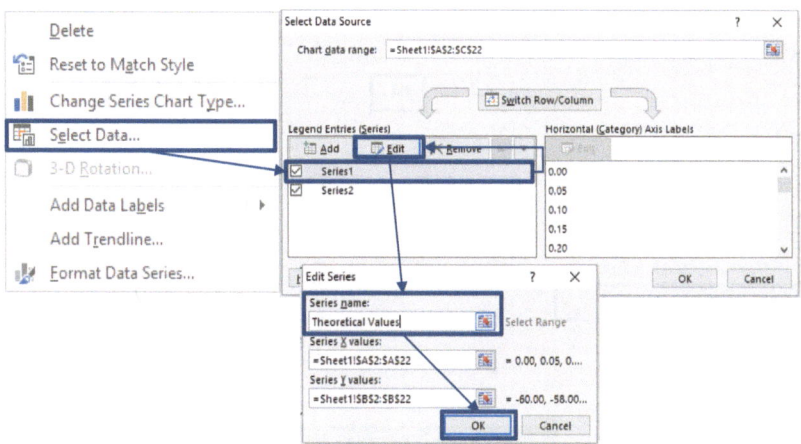

Fig. 2.16   Changing the legend

options for a visually appealing graph. Therefore, these values should be changed as necessary. This is shown in Fig. 2.17 (left and middle). Select the y-axis, open the `Format Axis` window, and select the `Axis Option` tab. Since we wish to change where the x-axis crosses the y-axis, we will select the `Axis value` for the `Horizontal axis crosses` to be -70. This will put the x-axis at the bottom of the chart. This is shown in Fig. 2.17 (left and right). The resulting graph is shown in Fig. 2.18.

Next, we need to label the axes. Select any element on the chart. Go to the `Chart Tools Ribbon, Design Option`. Go to the `Add Chart Element, Axis Titles, Primary Horizontal`. This is shown in Fig. 2.19. Click on the axis label and enter "Concentration of A, [A] (mol/ℓ)". Repeat the above steps but select `Primary Vertical` and enter "Freezing Point Temperature, T (°C)".

Finally, we should add a title to the chart. Click on the `Chart Title` label and enter "Freezing Point Experiment". If we are copying the figure to another document, such as Word, then it is best to simply delete the `Chart Title` textbox. The information will be provided in the obligatory caption.

The global font for the chart can be changed by selecting the outer border of the graph and setting the font to "Times New Roman" and size 12. All the characters will change to the new font and size.

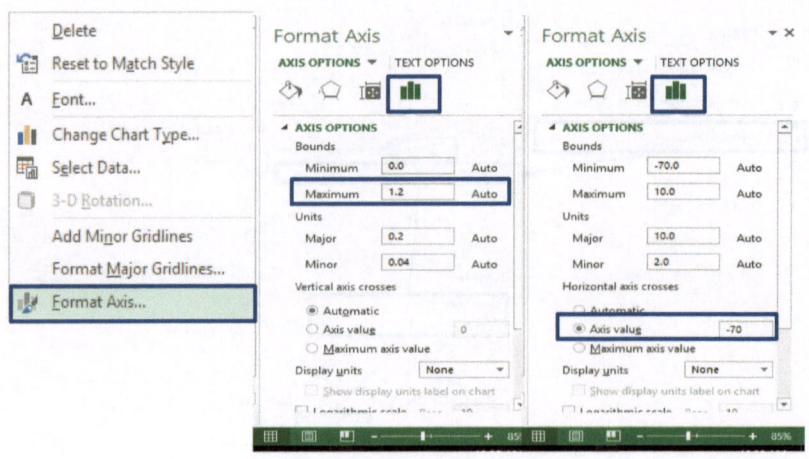

Fig. 2.17 Formatting the axes: (left) initial selection, (middle) formatting the x-axis, and (right) formatting the y-axis

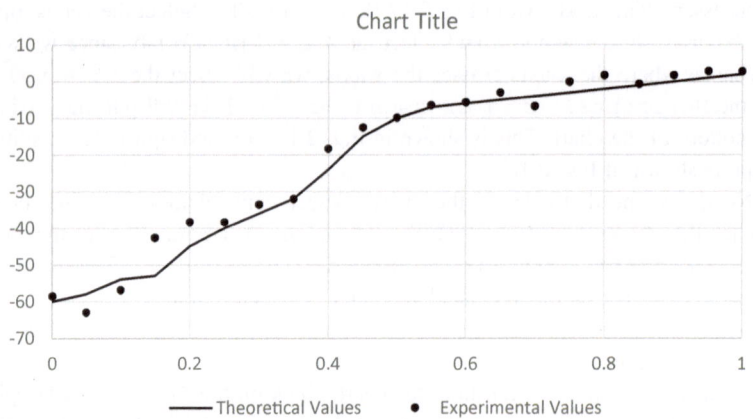

Fig. 2.18 Chart after formatting the legend and axes

The location of the chart can be changed by going to the Chart Tools Ribbon, Design Option and selecting the Move Chart icon. You can then select New sheet in the window that appears and give the new sheet a useful

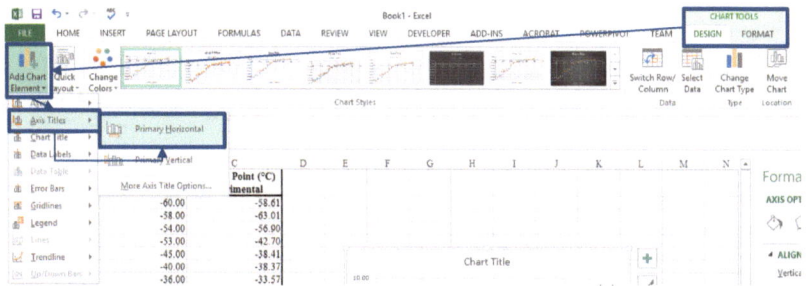

Fig. 2.19 Adding labels to the axes

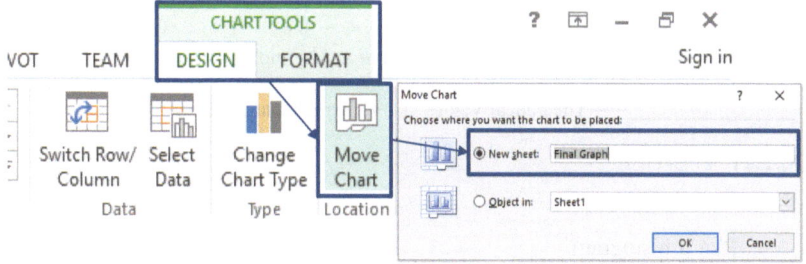

Fig. 2.20 Changing the location of a chart

name. This is shown in Fig. 2.20. It is better to copy the chart from a new sheet rather than as an object.

The final chart is shown in Fig. 2.21.

2.3   Practice Questions

Solve the following questions using Excel.

(1) For the data set {1, 3, 5, 2, 5, 7, 5, 2, 8, 5},

    a. Compute the mean, mode, and median.

    b. Compute the variance, median absolute difference, and range.

    c. Compute the first, second, and third quartiles.

    d. Plot a box-and-whisker plot.

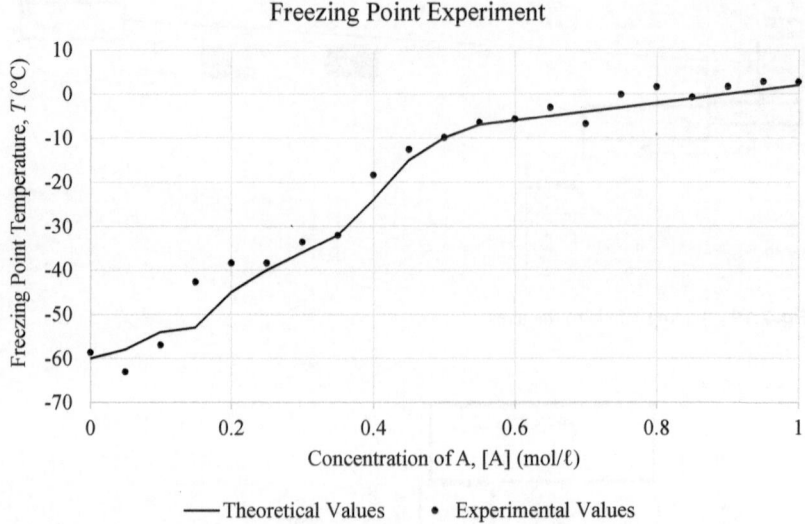

Fig. 2.21 Final chart

    e. Plot a histogram.

(2) Consider the data in Table 2.3 that shows the flow rate of steam in kg/h through a pipe. Due to the presence of stiction and other nonlinearities in the control valve, a new control algorithm is being proposed. The engineer in charge of making the change has to evaluate whether the new algorithm is better. A better algorithm is defined as one that reduces the variance of the steam flow rate and can keep the process closer to the desired setpoint of 8.5 kg/h. The original and new control methods are both tested for 2 h and the data are collected every 5 min. Plot the available data and analyse it. Without using any formal statistical tests, suggest whether the proposed control algorithm is better than the original, base case.

(3) Take any large data set that is of interest to you and analyse it using the methods presented in this chapter. The data set should have at least 1,000 data points and two variables. You can then use this data set in subsequent chapters to perform additional analysis.

Table 2.3 Steam control data with two different methods (for Question 2)

Time (min)		5	10	15	20	25	30	35	40	45	50	55	60
Base	1 h	8.5	8.7	8.4	8.6	8.2	8.7	8.9	8.5	8.5	8.4	8.3	8.6
	2 h	8.2	8.4	8.3	8.2	8.4	8.5	8.8	8.3	8.6	8.7	8.5	8.3
New	1 h	8.4	8.5	8.4	8.5	8.6	8.3	8.6	8.7	8.2	8.3	8.4	8.5
	2 h	8.5	8.6	8.4	8.3	8.4	8.6	8.7	8.5	8.5	8.5	8.3	8.4

Table 2.2

Theoretical Statistics, Distributions, Hypothesis Testing, and Confidence Intervals

3

This chapter presents material associated with Chap. 2: Theoretical Foundation for Statistical Analysis *of the book* Statistics for Chemical and Process Engineers: A Modern Approach.

3.1 Summary of Relevant Functions

3.1.1 Excel Functions for Statistical Distributions

In Excel, the different functions for a statistical distribution are built up using a common pattern that consists of two parts: the tag for the given statistical distribution and the abbreviation for the required property. The property abbreviations are: `inv` for the inverse function, `pdf` for the probability density function, and `cdf` for the cumulative distribution function. These two tags are combined by placing a period between them. Thus, for example, the inverse function for the F-distribution will be given as `f.inv`, since the tag for the F-distribution is `f`. Table 3.1 summarises the Excel functions for common statistical distributions with brief examples. The extremely commonly used standard normal distribution is obtained by setting the mean to zero and the standard deviation to one for the normal distribution.

© The Author(s), under exclusive license to Springer Nature Switzerland AG 2024   35
Y. A.W. Shardt, *Using Excel to Solve Statistical Problems: A Practical Guide to the Book "Statistics for Chemical and Process Engineers"*,
https://doi.org/10.1007/978-3-031-65449-7_3

Table 3.1 Common statistical distributions in Excel

Distribution	Command	Definition	Example	Output
Normal distribution (norm)	norm.inv(p, mean, std)	Inverse	=norm.inv(0.95,0,1)	1.6449
	norm.dist(x, mean, std, false)	Probability density function	=norm.pdf(4,0,1, false)	1.338e−04
Student's t-distribution (t)	t.inv(p, df)	Inverse	=t.inv(0.95,12)	1.7823
	t.dist(x, df, false)	Probability density function	=t.dist(4,12,false)	0.0016
	t.cdf(x, df)	Cumulative distribution function	=t.cdf(0.5,12)	0.6869
χ^2-distribution (chisq)	chisq.inv(p, df)	Inverse of χ^2	=chisq.inv(0.95,3)	7.8147
	chisq.dist(x, df, false)	Probability density function	=chisq.dist(4,3, false)	0.1080
F-distribution (f)	f.inv(p, df1, df2)	Inverse	=f.inv(0.95,3,5)	5.4095
	f.dist(x, df1, df2, false)	Probability density function	=f.pdf(4,3,5,false)	0.0354

(continued)

Table 3.1 (continued)

Distribution	Command	Definition	Example	Output
Binomial distribution (binom)	binom.inv(p, n, q)	Inverse of χ^2	=binom.inv(0.95, 50,0.2)	15
	binom.dist(k, n, q, false)	Probability density function	=binom.dist(4,50,0.2,false)	0.0128
Poisson distribution (poisson)	poisson.dist(k, 1,false)	Probability density function	=poisson.dist(1,0.2, false)	0.1637

The Excel tag for the given distribution is given in brackets. The following are common abbreviations: p is the probability, x is the critical value, and df is the degrees of freedom

3.1.2 Hypothesis Testing

In order to simplify matters, all the required probabilities are stated in terms of left probabilities, that is, from $-\infty$. This means that all the formulae can then be directly used in Excel. Let the parameter of interest be given as θ, the computed test statistic as $r_{computed}$, the critical value of the test statistic as r_{crit}, and the α-error as α. In general, we can distinguish between three cases depending on the form of the alternative hypothesis, H_1:

(1) **Case 1:** H_1: $\theta \neq \hat{\theta}$:
  a. **Symmetric Distribution:** $|r_{computed}| \geq r_{crit} = \texttt{inv}(1 - 0.5\alpha)$
  b. **Unsymmetric Distribution:** $\texttt{inv}(0.5\alpha) = rcrit, lower \leq r_{computed} \leq r_{crit, upper} = \texttt{inv}(1 - 0.5\alpha)$
(2) **Case 2:** H_1: $\theta < \hat{\theta}$: $\texttt{inv}(\alpha) = r_{crit} \leq r_{computed}$
(3) **Case 3:** H_1: $\theta > \hat{\theta}$: $r_{computed} \geq r_{crit} = \texttt{inv}(1 - \alpha)$

where $\texttt{inv}$ represents the appropriately selected inverse for the given statistical distribution.

In certain fields and applications, instead of computing the critical value of the test statistic, the analysis is performed using p-values. In such cases, the p-value corresponding to a computed test statistic, $r_{computed}$, can be determined for each of the cases as follows:

(1) **Case 1:** H_1: $\theta \neq \hat{\theta}$:
  a. **Symmetric Distribution:** $p_{computed} = 2 * \texttt{cdf}(-\texttt{abs}(\texttt{rcomputed})) \leq 0.5\alpha$
(2) **Case 2:** H_1: $\theta < \hat{\theta}$: $p_{computed} = \texttt{cdf}(\texttt{rcomputed}) \leq \alpha$
(3) **Case 3:** H_1: $\theta > \hat{\theta}$: $p_{computed} = 1\texttt{-cdf}(\texttt{rcomputed}) \leq \alpha$

where $\texttt{cdf}$ represents the appropriately selected cumulative distribution function for the given statistical distribution. This approach does not work for case 1 and an unsymmetric distribution.

3.1.3 Confidence Intervals

In general, the $100(1-\alpha)\%$ confidence intervals are computed for case 1 using the following general formula

$$\boxed{\hat{\theta} - r_{lower}\sigma_\theta \leq \theta \leq \hat{\theta} + r_{upper}\sigma_\theta} \tag{3.1}$$

where r_{lower} is the lower-bound critical value, r_{upper} is the upper-bound critical value, and σ_θ is the standard deviation of the parameter estimate. In Excel terms, this means that the confidence interval can be written as

θ_hat− σ_θ*abs(inv(0.5α)) $\leq$ θ $\leq$ θ_hat + σ_θ*inv(1−0.5α)

where inv represents the appropriately selected inverse for the given statistical distribution.

3.1.3.1 Common Tests and Their Corresponding Distributions

Table 3.2 presents a brief summary of the most common statistical tests and how they would be implemented.

Table 3.2 Summary of the common statistical tests and their required parameters. Let n be the number of data points

Test	Mathematical form	Distribution	Comment
Single mean	$t_{computed} = \dfrac{\hat{\mu}-\mu}{\hat{\sigma}/\sqrt{n}}$	Student's t-distribution with $n-1$ degrees of freedom	If $n > 30$ or the true standard deviation is known, then the standard normal distribution can be used
Single variance	$\chi^2_{computed} = \dfrac{(n-1)\hat{\sigma}^2}{\sigma^2}$	χ^2-distribution with $n-1$ degrees of freedom	
Two sample means	$Z_{computed} = \dfrac{\hat{\mu}_1-\hat{\mu}_2-\Delta}{\sqrt{\frac{\sigma_1^2}{n_1}+\frac{\sigma_2^2}{n_2}}}$	Standard normal distribution	Additional details regarding this complex case can be found in §2.7.6.1 of the textbook
Two sample variances	$F_{computed} = \dfrac{\hat{\sigma}_1^2}{\hat{\sigma}_2^2}$	F-distribution with n_1-1 and n_2-1 degrees of freedom	Most commonly used to test if $\sigma_1 > \sigma_2$

3.2   Detailed Examples

In this section, we will consider the two most common applications of statistics, hypothesis testing and confidence intervals, to answering questions about the mean and variance of a data sample. These methods will then be used in more details in subsequent chapters.

3.2.1   Testing a Mean Against a Theoretical Value

Consider the following data

$$\mathbb{X} = \{2.16, 2.71, 1.09, \ 0.40, 1.47, 1.13, 1.97\}$$

that are claimed to come from a normal distribution with a mean of 1. Set up and compute the required parameters for the hypothesis test that the two means are the same, compute the p-value, and compute the 95% confidence interval. Show that the results obtained are the same in all three cases. Set α equal to 0.05.

Before we can even solve the question, we will need to obtain the mean and standard deviation of the data set in Excel. First, enter the values into a row in Excel. Call this row `data`. Then enter into separate cells, the following formulae:

```
[mean] = average(data)                              = 1.5614
[std] = stdev(data)                                 = 0.7757
```

Since we are interested in determining if the computed mean is equal to the true mean, we will perform the following hypothesis test:

$$H_0 : \hat{\mu} = \mu$$

$$H_1 : \hat{\mu} \neq \mu$$

This corresponds to case 1. Since we have a small sample and do not know the true standard deviation, we will have to use the Student's t-distribution, which is a symmetric distribution. Therefore, from the first row of Table 3.1, the computed value of the test statistic can be found in Excel as

```
[tcomp] =(mean-1)/(std/sqrt(count(data)))           = 1.9150
```

The critical value can be computed using the case 1 symmetric formula to give

```
[tcrit] =t.inv(1-0.5*0.05,count(data)-1)          = 2.4469
```

Note that we need to include the degrees of freedom in the formula for the inverse.

We can now compare the two values to determine if the alternative hypothesis is in fact correct. In Excel, we can write this comparison as

```
=abs(tcomp)>=tcrit                                = FALSE
```

The answer is noted as being FALSE. This means that the alternative hypothesis can be rejected and the null hypothesis may be correct, that is, the data sample has the given mean of one.

The p-value corresponding to the given computed critical value can be computed using the formula for a symmetric case 1, which gives in Excel:

```
[p] =2*t.dist(-abs(tcomp),count(data)-1,true)     = 0.1040
```

Comparing the computed p-value against the threshold value of 0.5α gives

```
=p<=0.5*0.05                                      =FALSE
```

which once again confirms that the alternative hypothesis can be rejected. There should never be any difference in the outcomes between the different approaches.

Finally, let us consider the 95% confidence interval for the given parameter estimate. The formula for a confidence interval requires that we determine the upper and lower bounds using an appropriate statistical distribution. In our case, the appropriate distribution, which can be found from Table 3.1, is the Student's t-distribution. The standard deviation of the parameter is simply $\hat{\sigma}/\sqrt{n}$. Thus, in Excel, the upper and lower limits can be found as follows:

```
[rl] =mean-std/sqrt(count(data))*abs(tcrit)       =0.8440
[ru] =mean+std/sqrt(count(data))*abs(tcrit)       =2.2788
```

Thus, the confidence interval can be written as [0.8440, 2.2788]. Since the true value of one lies within the given confidence interval, it can be concluded that the confidence interval covers the true value. This implies that the computed mean and the true mean could be the same, which agrees with our previous two approaches. Normally, we would select one of these approaches and use them throughout a given problem.

3.2.1.1 Comparing Two Sample Variances

As a plant engineer, you are testing a new drying procedure for the plant. Two options exist: A and B. Option A is currently in place, while option B is a faster new method. It is desired to determine if the variance of option B is greater than that of option A in which case option B should not be selected.

Option A gave the following product quality: 95.6, 97.3, 95.6, 95.4, 99.4, 97.2, 92.2, 92.8, 94.3, and 92.6. Option B gave the following product quality: 89.2, 94.2, 93.9, 93.2, 94.7, 91.7, 93.2, 92.4, 91.8, and 91.5.

Solve this problem using hypothesis tests by computing the critical test statistic and also the p-value.

In order to solve this problem in Excel, it is first necessary to place the data in two rows (or columns) in Excel. For the sake of the discussion, it will be assumed that data for Option A has been placed in column A and the data for Option B in column B. Column A has been labelled as `opt_A` and Column B as `opt_B`. Then, the standard deviation can be computed as follows

```
[sig_A]=stdev.s(opt_A)                            =2.3268
[sig_B]=stdev.s(opt_B)                            =1.6206
[n_A]=count(opt_A)                                   =10
[n_B]=count(opt_B)                                   =10
```

The formal statistical test can be stated as.

$$H_0 : \sigma_B = \sigma_A$$

$$H_1 : \sigma_B > \sigma_A$$

Note that the order in which the two variances are listed is important so that the correct ratio is formed. From here, we see that we have a case 3 example (this is denoted as case 2 in the textbook due to the fact that $\sigma_2 > \sigma_1$ is the same as $\sigma_1 < \sigma_2$). As well, since we are comparing two sample variances, we will need to look at the last row in Table 3.2. This shows that we will need to use the F-distribution and that the test statistic is simply a ratio between the two variances. Implementing this in Excel gives

Option A	Option B
95.6	89.2
97.3	94.2
95.6	93.9
95.4	93.2
99.4	94.7
97.2	91.7
92.2	93.2
92.8	92.4
94.3	91.8
92.6	91.5

Option	σ	n
A	2.326753	10
B	1.620562	10

F_{comp}	0.696491
F_{crit}	3.178893
Result	FALSE

p-value	0.700671
critical value	0.05
Result	FALSE

Option	σ	n
A	=STDEV.S(opt_A)	=COUNT(opt_A)
B	=STDEV.S(opt_B)	=COUNT(opt_B)

F_{comp}	=sig_B/sig_A
F_{crit}	=F.INV(1-0.05,n_B-1,n_A-1)
Result	=fcomp>=fcrit

p-value	=1-F.DIST(fcomp,n_B-1,n_A-1,TRUE)
critical value	0.05
Result	=pvalue<=E11

Fig. 3.1 (Left) Numerical results and (right) the corresponding Excel formulae

```
[fcomp]=sig_B/sig_A                               =0.6965
[fcrit]=f.inv(1-0.05,n_B-1,n_A-1)                 =3.1789
```

Thus, the conclusion can be determined by comparing the two values to give

```
=fcomp >=fcrit                                    =FALSE
```

This means that the alternative hypothesis can be rejected and the null hypothesis may be correct, that is, the two data samples have the same variance.

The corresponding p-value can be computed using the case 3 formula to give

```
[pvalue] =1-f.dist(fcomp,n_B-1,n_A-1,TRUE)        =0.7007
```

Comparing against a value of $\alpha = 0.05$ gives

```
=pvalue<=0.05                                     =FALSE
```

which once again confirms that the alternative hypothesis can be rejected. There should never be any difference in the outcomes between the different approaches. An implementation of the above procedure in an Excel spreadsheet is shown in Fig. 3.1.

3.3 Practice Questions

Solve the following questions using Excel.

(1) For the following experiments, compute the 95% confidence intervals and determine whether the data comes from the stated population.
   a. $\mathbb{X} = \{3, 2.3, 4.5, 1.2, 5.6, 2.3, 4.5\}$, $\mu = 3$.
   b. $\hat{\mu} = 4.2$, $\hat{\sigma} = 1.2$, $n = 100$, $\mu = 3$.
   c. $\hat{\mu} = 0.2$, $\hat{\sigma} = 5$, $n = 10$, $\mu = 3$.
   d. $\hat{\sigma} = 1.5$, $n = 10$, $\sigma = 5$.
(2) Verify the central limit theorem for the following distributions: normal, χ^2-, and F-distributions. Compute the mean value for multiple samplings of these distributions. Do they converge to the normal distribution?

Regression Analysis and Design of Experiments

4

This chapter presents material associated with Chap. 3: Regression *and* Chapter 4: Design of Experiments *of the book* Statistics for Chemical and Process Engineers: A Modern Approach.

4.1 Summary of Relevant Functions

4.1.1 Excel Formulae for Ordinary, Linear Regression

Consider a regression model given by

$$y = \sum_{i=1}^{n} \beta_i f_i(\vec{x}) + \varepsilon = \vec{a}\vec{\beta} + \varepsilon \qquad (4.1)$$

with n regressors and m data points (that is, different values of the regressors and the corresponding outputs), define the following vectors and matrices

$$\mathcal{A} = \begin{bmatrix} f_1(\vec{x}_1) & f_2(\vec{x}_1) & \cdots & f_n(\vec{x}_1) \\ f_1(\vec{x}_2) & f_2(\vec{x}_2) & \cdots & f_n(\vec{x}_2) \\ \vdots & & \ddots & \vdots \\ f_1(\vec{x}_m) & f_2(\vec{x}_m) & \cdots & f_n(\vec{x}_m) \end{bmatrix} \qquad (4.2)$$

$$\vec{\beta} = <\beta_1, \beta_2, \dots, \beta_n >^T \qquad (4.3)$$

© The Author(s), under exclusive license to Springer Nature Switzerland AG 2024 45
Y. A.W. Shardt, *Using Excel to Solve Statistical Problems: A Practical Guide to the Book "Statistics for Chemical and Process Engineers"*,
https://doi.org/10.1007/978-3-031-65449-7_4

$$\vec{y} = < y_1, y_2, \ldots, y_m >^T \tag{4.4}$$

$$\vec{a}_{\vec{x}_d} = < f_1(\vec{x}_d), f_2(\vec{x}_d), \ldots, f_n(\vec{x}_d) > \tag{4.5}$$

4.1.1.1 Solution Without Using the Excel Template

Table 4.1 summarises the Excel formulae for performing linear regression without using any special Excel templates.

4.1.1.2 Solution Using the Excel Template

Although it is possible to use the Excel to create a regression solution every time using the formulae presented in Table 4.1, it is easier to use a predefined Excel template file that has most of the formulae already implemented and ready to use. The Excel template file is called linearregression.xltm and can be found on the book website. A screenshot of the plain template is shown in Fig. 4.1. The yellow blocks are where the required data are entered. The green block represents the row in which an array formula needs to be entered. The complete green row should be selected and then the first cell highlighted. Finally, press Ctrl + Shift + Enter to copy the array formula to the entire green row. Adding additional parameters and data points will also require that the formulae be appropriately copied down. The spreadsheet automatically creates the normal probability plot for the residuals and plots of the residuals as a function of y and $\hat{y}$, as well as a time-series plot of the residuals. Additional plots can be created by the user. This template requires that the internal macros be enabled. As well, the array formulae need to be properly entered.

4.1.2 Excel Formulae for Weighted Linear Regression

Consider a regression model given by

$$y = \sum_{i=1}^{n} \beta_i f_i(\vec{x}) + w^{-1}\varepsilon = \vec{a}\vec{\beta} + w^{-1}\varepsilon \tag{4.6}$$

with n regressors and m data points (that is, different values of the regressors and the corresponding outputs), define the following vectors and matrices

Table 4.1 Excel formulae for ordinary, linear regression by hand

Name	Mathematical formula	Excel equivalent (with matrix size)
Fisher information matrix	$\mathcal{F} = \left(A^T A\right)^{-1}$	`[F] = MINVERSE(MMULT(TRANSPOSE(A),A))` Size: $[n, n]$
Parameter estimates	$\hat{\beta} = \left(A^T A\right)^{-1} A^T \vec{y}$	`[beta] = MMULT(MMULT(F,TRANSPOSE(A)),y)` Size: $[n, 1]$
Standard deviation of the model	$\hat{\sigma} = \sqrt{\dfrac{\vec{y}^T \vec{y} - \hat{\beta}^T A^T \vec{y}}{m-n}}$	`[sig_mod] = SQRT((MMULT(TRANSPOSE(y),y)-MMULT(MMULT(TRANSPOSE(beta),TRANSPOSE(A)),y))/(m-n))` Size: $[1, 1]$
Residuals	$\vec{\varepsilon} = \vec{y} - A\hat{\beta}$	`[res] = y-MMULT(A,beta)` Size: $[m, 1]$
$100(1-\alpha)\%$ confidence interval for β_i	$\hat{\beta}_i \pm t_{1-\frac{\alpha}{2}, m-n} \hat{\sigma} \sqrt{(A^T A)^{-1}_{ii}}$	`[Delta] = T.INV(1-alpha/2, m-n) *sig_mod*sqrt(F(i, i))`[1] Size: `[1, 1]`
Predicted value	$\hat{y} = \vec{a}_{x_d}^T \hat{\beta}$	`[y_hat] = axd*beta` Size: $[1, 1]$
$100(1-\alpha)\%$ mean response confidence intervals	$\hat{y} \pm t_{1-\frac{\alpha}{2}, m-n} \hat{\sigma} \sqrt{\vec{a}_{x_d}(A^T A)^{-1} \vec{a}_{x_d}^T}$	`[Delta] = T.INV(1-alpha/2,m-n) * sig_mod* SQRT(MMULT(MMULT(axd,F),TRANSPOSE(axd)))` Size: $[1, 1]$

(continued)

---

[1] You will need to manually select the cell corresponding to the required entry in the Fisher information matrix.

Table 4.1 (continued)

Name	Mathematical formula	Excel equivalent (with matrix size)
$100(1-\alpha)\%$ predictive confidence intervals	$\hat{y} \pm t_{1-\frac{\alpha}{2}, m-n}\, \hat{\sigma} \sqrt{1 + \vec{a}_{x_d}(A^T A)^{-1}\vec{a}_{x_d}^T}$	`[Delta] = T.INV(1-alpha/2,m-n) * sig_mod*SQRT(1 + MMULT(MMULT(axd,F),TRANSPOSE(axd)))` Size: [1, 1]
SSR	$SSR = \sum (\hat{y}_i - \bar{y})^2$	`[SSR] =` `SUM((MMULT(A,beta)-AVERAGE(y))^2)`
SSE	$SSE = \sum (y_i - \hat{y}_i)^2 = \vec{\varepsilon}^T \vec{\varepsilon}$	`[SSE] = SUM(res^2)`
TSS	$TSS = \sum (y_i - \bar{y})^2$	`[TSS] = SSR + SSE` `[TSS] = SUM((y-AVERAGE(y))^2)`
F-statistic[2]	$F = \dfrac{SSR_{/k}}{SSE_{/m-n}}$	`[Fstat] = (SSR/k)/(SSE/(m-n))`
F-critical	$F_{1-\alpha,\, k,\, m-n}$	`[Fcrit] = F.INV(1-alpha, k, m-n)`
R^2	$R^2 = \frac{SSR}{TSS} = 1 - \frac{SSE}{TSS}$	`[Rsq] = SSR/TSS` `[Rsq] = 1-SSE/TSS`
R^2_{adj}	$R^2_{adj} = 1 - (1-R^2)\left(\frac{m-1}{m-n}\right)$	`[R2adj] = 1-(1-Rsq) * (m-1)/(m-n)`

Let A equal $\mathcal{A}$ y equal $\hat{y}$, and axd equal $\vec{a}_{x_d}$. Note that all formulae are array formulae and require that the appropriately sized array be selected and CTRL + SHIFT + ENTER be pressed

---

[2] For most applications, $k = n - 1$. If the model has no constant term (β_0), that is, there is no column of ones in the $\mathcal{A}$-matrix, then $k = n$.

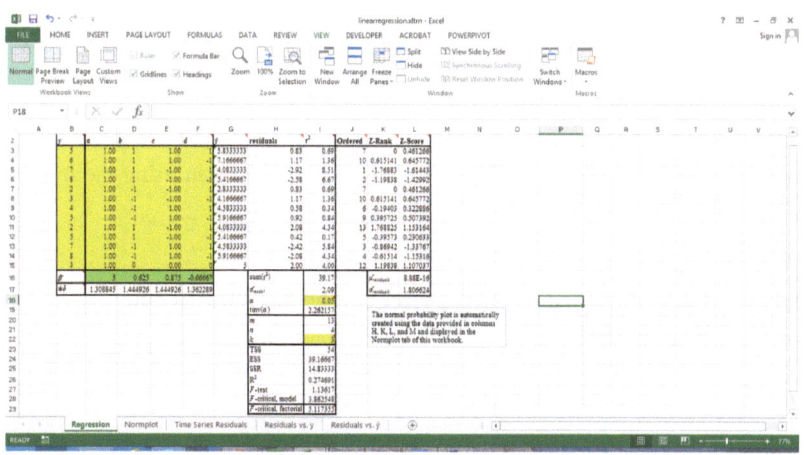

Fig. 4.1 Linear regression template

$$\mathcal{A} = \begin{bmatrix} f_1(\vec{x}_1) & f_2(\vec{x}_1) & \cdots & f_n(\vec{x}_1) \\ f_1(\vec{x}_2) & f_2(\vec{x}_2) & \cdots & f_n(\vec{x}_2) \\ \vdots & & \ddots & \vdots \\ f_1(\vec{x}_m) & f_2(\vec{x}_m) & \cdots & f_n(\vec{x}_m) \end{bmatrix} \tag{4.7}$$

$$\vec{\beta} = <\beta_1, \beta_2, \ldots, \beta_n>^T \tag{4.8}$$

$$\vec{y} = <y_1, y_2, \ldots, y_m>^T \tag{4.9}$$

$$\vec{a}_{\vec{x}_d} = <f_1(\vec{x}_d), f_2(\vec{x}_d), \ldots, f_n(\vec{x}_d)> \tag{4.10}$$

$$\mathcal{W} = \begin{bmatrix} w_1 & 0 & \cdots & 0 \\ 0 & w_2 & 0 & 0 \\ & & \ddots & \\ 0 & 0 & \ddots & 0 \\ 0 & \cdots & 0 & w_m \end{bmatrix} \tag{4.11}$$

4.1.2.1 Solution Using Excel Functions

Table 4.2 summarises the Excel formulae for performing weighted, least-squares, linear regression. Unfortunately, Excel does not have a built-in function that can compute the square root of a matrix. If we only allow diagonal weighting matrices, then the square root of the matrix is equal to the square root of each individual component, which is the result in Excel when using the formula = W^0.5.

4.1.3 Excel Formulae and Procedures for Nonlinear Regression

Let the nonlinear regression problem be written as

$$y = g\left(\vec{\beta}, \vec{x}, \varepsilon\right) \tag{4.12}$$

where the optimisation problem is

$$\min_{\vec{\beta}} \sum_{i=1}^{m} \left(y_i - g\left(\vec{\beta}, \vec{x}_i, \varepsilon_i\right)\right)^2 \tag{4.13}$$

All nonlinear regression approaches use numerical methods, such as the Gauss–Newton or Levenberg–Marquardt algorithms, to search for the optimal point. The derivative matrix of this problem, called the grand Jacobian matrix, $\mathcal{J}$, plays a role similar to that of the $\mathcal{A}$ matrix in linear regression. The Jacobian, $\mathcal{J}'$, for the system can be calculated as

$$\mathcal{J}' = \left[\frac{\partial g\left(\vec{\beta}, \vec{x}, \varepsilon\right)}{\partial \beta_1} \quad \frac{\partial g\left(\vec{\beta}, \vec{x}, \varepsilon\right)}{\partial \beta_2} \quad \cdots \quad \frac{\partial g\left(\vec{\beta}, \vec{x}, \varepsilon\right)}{\partial \beta_n} \right] \tag{4.14}$$

The value of $\mathcal{J}'$ is determined for each of the data points present to obtain the grand Jacobian matrix, $\mathcal{J}$. Thus, $\mathcal{J}$ can be written as

Table 4.2 Excel formulae for weighted linear regression

Name	Mathematical formula	Excel equivalent
Fisher information matrix	$\mathcal{F} = (A^T \mathcal{W} A)^{-1}$	[F] = MINVERSE(MMULT(MMULT(TRANSPOSE(A),W),A)) Size: $[n, n]$
Parameter estimates	$\hat{\beta} = (A^T \mathcal{W} A)^{-1} A^T \mathcal{W}\vec{y}$	[beta] = MMULT(MMULT(MMULT(F,TRANSPOSE(A)),W),y) Size: $[n, 1]$
Standard deviation of the model	$\hat{\sigma} = \sqrt{\dfrac{\vec{y}^T \mathcal{W}\vec{y} - \hat{\beta}^T A^T \mathcal{W}\vec{y}}{m-n}}$	[sig_mod] = SQRT((MMULT(MMULT(TRANSPOSE(y),W),y)-MMULT(MMULT(MMULT(TRANSPOSE(beta),TRANSPOSE(A)),W),y))/(m-n)) Size: $[1, 1]$
Residuals	$\vec{\varepsilon} = \mathcal{W}^{0.5}\left(\vec{y} - A\hat{\beta}\right)$	[res] = MMULT(W^0.5,y-MMULT(A,beta)) Size: $[m, 1]$
$100(1-\alpha)\%$ confidence interval for β_i	$\hat{\beta}_i \pm t_{1-\frac{\alpha}{2},m-n}\hat{\sigma}\sqrt{(A^T \mathcal{W} A)_{ii}^{-1}}$	[Delta] = T.INV(1-alpha/2,m-n) *sig_mod*SQRT(F(i,i))[3] Size: [1, 1]
Predicted value	$\hat{y} = \vec{a}_{x_d}^T\hat{\beta}$	[y_hat] = axd*beta Size: [1, 1]
$100(1-\alpha)\%$ mean response confidence intervals	$\hat{y} \pm t_{1-\frac{\alpha}{2},m-n}\hat{\sigma}\sqrt{\vec{a}_{x_d}(A^T \mathcal{W} A)^{-1}\vec{a}_{x_d}^T}$	[Delta] = T.INV(1-alpha/2,m-n)*sig_mod*SQRT(MMULT(MMULT(axd,F),TRANSPOSE(axd))) Size: [1, 1]

(continued)

[3] You will need to manually select the cell corresponding to the required entry in the Fisher information matrix.

Table 4.2 (continued)

Name	Mathematical formula	Excel equivalent
$100(1-\alpha)\%$ predictive confidence intervals	$\hat{y} \pm t_{1-\frac{\alpha}{2}, m-n}\hat{\sigma}\sqrt{\dfrac{1}{w_0} + \vec{a}_{\vec{x}_d}(A^T A)^{-1}\vec{a}_{\vec{x}_d}^T}$	`[Delta] = T.INV(1-alpha/2, m-n-n_` `sigma)*sig_mod* sqrt(1/w0 +` `MMULT(MMULT(axd,F),` `TRANSPOSE(axd))` Size: [1, 1]
SSR	$SSR = \sum w_i(\hat{y}_i - \bar{y})^2$	`[SSR] = SUM(MMULT(W,` `(MMULT(A,beta)-AVERAGE(y))^2)`
SSE	$SSE = \sum w_i(y_i - \hat{y}_i)^2 = \vec{\varepsilon}^T\vec{\varepsilon}$	`SSE = SUM(res^2)`
TSS	$TSS = \sum w_i(y_i - \bar{y})^2$	`TSS = SSR + SSE`
F-statistic[4]	$F = \dfrac{SSR/k}{SSE/m - n}$	`[Fstat] = (SSR/k)/(SSE/(m-n))`
F-critical	$F_{1-\alpha, k, m-n}$	`[Fcrit] = F.INV(1-alpha,k, m-n)`
R^2	$R^2 = \dfrac{SSR}{TSS} = 1 - \dfrac{SSE}{TSS}$	`[Rsq] = SSR/TSS` `[Rsq] = 1-SSE/TSS`
R^2_{adj}	$R^2_{adj} = 1 - (1 - R^2)\left(\dfrac{m-1}{m-n}\right)$	`[R2adj] = 1-(1-Rsq) * (m-1)/(m-n)`

Let A equal $\mathcal{A}$, W equal $\mathcal{W}$, y equal $\vec{y}$, and axd equal $\vec{a}_{\vec{x}_d}$. Note that all formulae are array formulae and require that the appropriately sized array be selected and `CTRL` + `SHIFT` + `ENTER` be pressed

---

[4] For most applications, $k = n - 1$. If the model has no constant term (β_0), that is, there is no column of ones in the $\mathcal{A}$-matrix, then $k = n$.

$$
\mathcal{J} = \begin{bmatrix} \mathcal{J}_1^{\cdot} \\ \mathcal{J}_2^{\cdot} \\ \vdots \\ \mathcal{J}_m^{\cdot} \end{bmatrix} = \begin{bmatrix} \frac{\partial g\left(\vec{\beta},\vec{x}_1,\varepsilon\right)}{\partial \beta_1} & \frac{\partial g\left(\vec{\beta},\vec{x}_1,\varepsilon\right)}{\partial \beta_2} & \cdots & \frac{\partial g\left(\vec{\beta},\vec{x}_1,\varepsilon\right)}{\partial \beta_n} \\ \frac{\partial g\left(\vec{\beta},\vec{x}_2,\varepsilon\right)}{\partial \beta_1} & \frac{\partial g\left(\vec{\beta},\vec{x}_2,\varepsilon\right)}{\partial \beta_2} & \cdots & \frac{\partial g\left(\vec{\beta},\vec{x}_2,\varepsilon\right)}{\partial \beta_n} \\ \vdots & \vdots & & \vdots \\ \frac{\partial g\left(\vec{\beta},\vec{x}_m,\varepsilon\right)}{\partial \beta_1} & \frac{\partial g\left(\vec{\beta},\vec{x}_m,\varepsilon\right)}{\partial \beta_2} & \cdots & \frac{\partial g\left(\vec{\beta},\vec{x}_m,\varepsilon\right)}{\partial \beta_n} \end{bmatrix} \quad (4.15)
$$

Since it is necessary to solve the problem using Excel's built-in Solver, it makes sense to simply present the solution using the nonlinear regression template. Without this template, it would be necessary to set everything up similarly and then use the Solver to find the solution. In all cases, VBA scripts will need to be written to provide the derivatives and perhaps also the function values themselves. The template is called `nonlinearregression.xltm` and can be downloaded from the book website. A screenshot of the plain template is shown in Fig. 4.2. The yellow blocks are where the required data are entered. Note that Solver needs to be used to obtain a solution to the problem. The configuration of Solver is shown as an inset in Fig. 4.2. The layout and formatting of the results are similar to the linear regression case. Two important differences are that the model and its Jacobian must be entered as a macro and that Solver must be used. The spreadsheet automatically creates the normal probability plot for the residuals and plots of the residuals as a function of y and $\hat{y}$, as well as a time-series plot of the residuals. Additional plots can be created by the user.

The template comes with four predefined functions for creating the model and the corresponding Jacobian. Each function takes the same inputs: the range corresponding to the parameters and the range corresponding to the inputs. The four functions are: `model`, `dydb1`, `dydb2`, and `dydb3`. This template requires that the internal macros be enabled and Solver installed.

4.1.4   Excel Formulae for Design of Experiments

Since the analysis and implementation of design of experiments is mostly performed using the same methods as linear regression, there are not many new formulae to consider. Table 4.3 summarise the new formulae for design of experiments.

As well, an Excel template for factorial analysis has been created. It has the same format as the linear-regression template, but contains the information relevant to factorial design. The file is called

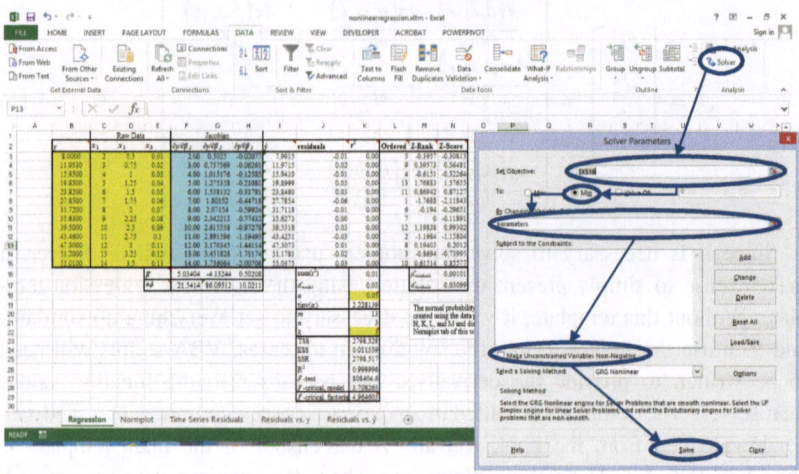

Fig. 4.2 Nonlinear regression template. The inset shows how to set up the Solver (Excel 2019)

Table 4.3 Excel formulae for design of experiments

Name	Mathematical formula	Excel equivalent
Sum of squares due to a regressor	$SSR_i = \left(\overline{A}^T \overline{A}\right)_{ii} \hat{\beta}_i^2$	`[Fr] = MMULT(TRANSPOSE(A,A))` `[SSRi] = Fr(i,i)*beta(i)^2`
F-statistic for a parameter	$F_i = \dfrac{SSR_i}{\frac{SSE}{l^k(n_R-1)}}$	`[Fi] = (SSRi)/(SSE/1^k/(nR-1))`
F-critical for a parameter	$F(1-\alpha, 1, l^k(n_R-1))$	`[Fcritp] = F.INV(1-alpha,1, 1^k*(nR-1))`

`factorialdesigntemplate.xltm` and can be downloaded from the book website. A screenshot of the plain template is shown in Fig. 4.3. The yellow blocks are where the required data are entered. The green block represents the row in which an array formula needs to be entered. The complete green row should be selected and then the first cell highlighted. Finally, press `Ctrl + Shift + Enter` to copy the array formula to the entire green row. Adding additional

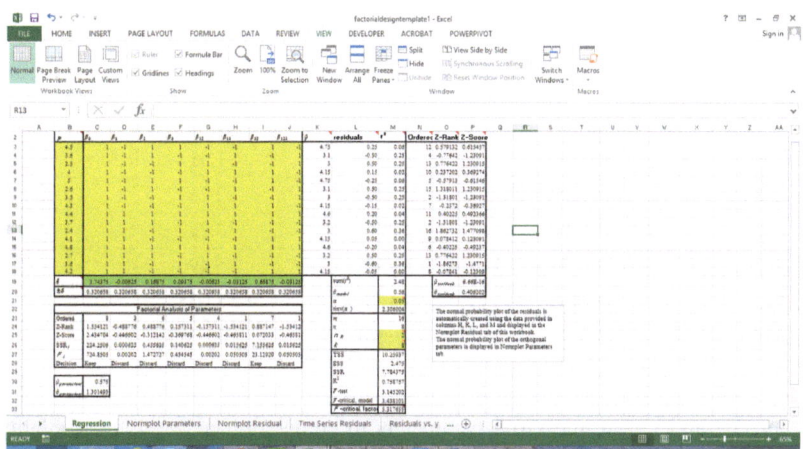

Fig. 4.3   Analysis of factorial experiments template

parameters and data points will also require that the formulae be appropriately copied down.

The spreadsheet automatically creates the normal probability plot for the parameters and residuals as well as plots of the residuals as a function of y and $\hat{y}$ and a time-series plot of the residuals. Additional plots can be created by the user. This template requires that the internal macros be enabled. As well, the array formulae need to be properly entered.

4.2   Detailed Examples

4.2.1   Linear Regression

Consider the problem of obtaining the values of the parameters in a theoretical equation that describes the osmotic pressure of the sodium chloride (NaCl) salt and hydroxyethyl starch (HES, chemical formula $(C_6H_{10}O_5)_m(C_2H_5O)_n$). Based on the virial equation of state, it is assumed that the following equation can be used to describe the osmolality (Π) of such a mixture

$$\Pi = B_3 m_3^2 + B_3 k_{diss} m_2 m_3 + C_3 m_3^3 + k_c \qquad (4.16)$$

Table 4.4 Fitting the virial equation

m_2 (millimol/kg solv)	m_3 (millimol/kg solv)	k_c (milliosm/kg solv)	Π (milliosm/kg solv)
0	0.0000	0	0
600	0.0390	1,052	1,314
1,268	0.0823	2,326	2,267
2,013	0.1307	3,879	3,712
2,852	0.1852	5,792	5,496
3,803	0.2469	8,170	8,035
4,889	0.3175	11,161	11,513

where B_3 and C_3 are the virial parameters to be determined, m_2 is the molality of NaCl in millimol/kg of solvent, m_3 is the molality of HES in millimol/kg of solvent, k_{diss} is the disassociation constant that is equal to 1.678, and k_c is a known constant that depends on the system being analysed. An experiment was run where the ratio of the mass of HES to the mass of NaCl was fixed to 0.5. The results obtained are shown in Table 4.4. Data used for this example come from Prickett et al. (2011).

Before linear regression can be applied, the above equation must be rearranged so that all known constant information is on the left-hand side and all the unknown variables are on the right-hand side. Thus, the equation would be rewritten as

$$\Pi - k_c = B_3\left(m_3^2 + k_{diss}m_2m_3\right) + C_3m_3^3 \qquad (4.17)$$

The required variables would be defined as

$$y = \Pi - k_c$$
$$\vec{x} = \left\langle m_3^2 + k_{diss}m_2m_3, m_3^3\right\rangle \qquad (4.18)$$
$$\vec{\beta} = \langle B_3, C_3\rangle^T$$

In order to solve this example in Excel, we will use the linear-regression template that has been developed. Therefore, open the file linearregression.xltm and accept any macros. We will see that the template assumes that there are four parameters to be estimated. However, in our example, we only have two. Therefore, we will delete columns D and E. It is always best to delete (add) interior columns as this will automatically update the variables to include the new situation, for example, in our case, the A-range will change from being defined as C3:F15 to C3:D5. The same will happen with the parameter range beta. As well, we need to make sure that we have sufficient rows for all our data. The original template contains 13 rows, but we only have 7 data points. Therefore, we need to remove 6 rows. For the purposes of this, we will remove the interior rows 6, 7, 8, 9, 10, and 11. Since we wish to have easy access to the original data, it will be placed in columns L, M, N, and O in the same order as Table 4.4. The data point is placed in L3 corresponding to a value of $m_2 = 0$.

Once this has been completed, we can enter the following formulae: in cell B3, the formula for computing the output

$$=O3-N3 \qquad\qquad\qquad =0$$

in cell C3, the formula for computing the regressor corresponding to B_3

$$=M3^2+L3*M3*1.678 \qquad\qquad\qquad =0$$

in cell D3, the formula for computing the regressor corresponding to C_3

$$=M3^2 \qquad\qquad\qquad =0$$

The three cells B3:D3 are selected and dragged down to fill the remaining yellow region (to B9:D9). If everything has been done properly, the values will be updated automatically and it should match the yellow region shown in Fig. 4.4.

The last thing that needs to be done is to compute the parameter values (green region). Select the green region (C10:D10), starting from C10. Place the cursor in the formula window, which should contain the formula for computing the parameters. Press CTRL + SHIFT + ENTER. The green region should be filled with the parameter estimates and all the other values should be updated automatically. The final thing that needs to be set is the k value in cell G16 (also in yellow). In our case, it should be set equal to G15, the number of parameters. The results are shown in Figs. 4.4 and 4.5.

Using the original data shows that the second point ($\Pi = 1{,}314$) is potentially an outlier, since its residual is extremely large. Thus, the row corresponding to this point (row 4 in the original layout) was deleted and the regression analysis was redone. The results are shown in Fig. 4.6. The results are much better as there

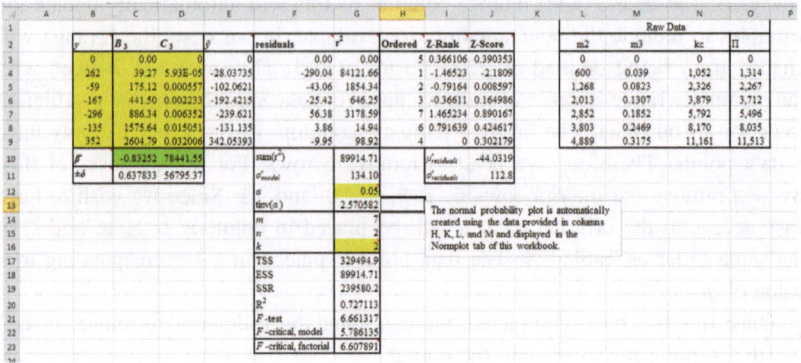

	y	B_3	C_3	$\hat{y}$	residuals	r^2	Ordered	Z-Rank	Z-Score		m2	m3	kc	Π
	0	0.00	0	0	0.00	0.00	5	0.366106	0.390353		0	0	0	0
	262	39.27	5.93E-05	-28.03735	-290.04	84121.66	1	-1.46523	-2.1809		600	0.039	1,052	1,314
	-59	175.12	0.000557	-102.0621	-43.06	1854.34	2	-0.79164	0.008597		1,268	0.0823	2,326	2,267
	-167	441.50	0.002233	-192.4215	-25.42	646.25	3	-0.36611	0.164986		2,013	0.1307	3,879	3,712
	-296	886.34	0.006352	-239.621	56.38	3178.59	7	1.465234	0.890167		2,852	0.1852	5,792	5,496
	-135	1575.64	0.015051	-131.135	3.86	14.94	6	0.791639	0.424617		3,803	0.2469	8,170	8,035
	352	2604.79	0.032006	342.05393	-9.95	98.92	4	0	0.302179		4,889	0.3175	11,161	11,513

$\hat{\beta}$	-0.83252	78441.55		sum(r^2)	89914.71	$\mu_{residual}$	-44.0319
$\pm\delta$	0.637833	56795.37		σ_{model}	134.10	$\sigma_{residual}$	112.8
				α	0.05		
				tinv(α)	2.570582		
				m	7		
				n	2		
				k	2		
				TSS	329494.9		
				ESS	89914.71		
				SSR	239580.2		
				R^2	0.727113		
				F-test	6.661317		
				F-critical, model	5.786135		
				F-critical, factorial	6.607891		

The normal probability plot is automatically created using the data provided in columns H, K, L, and M of this workbook. The Normplot tab of this workbook.

Fig. 4.4 Linear regression example: data analysis results

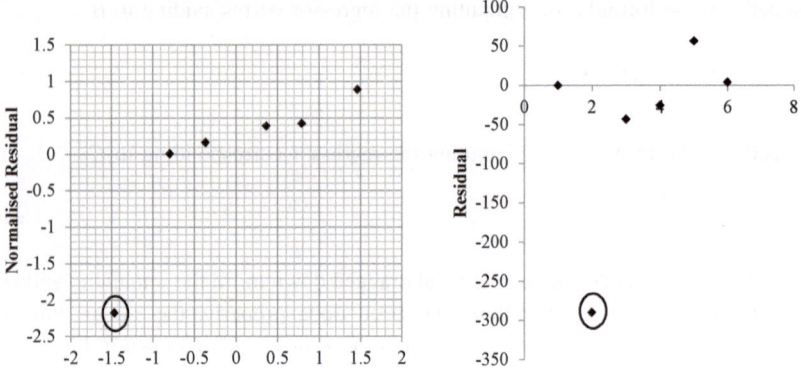

Fig. 4.5 Linear regression example: (left) normal probability and (right) time-series plots. The circled point is a potential outlier

are now no clear outliers and the data confidence intervals, especially for C_3, are much smaller. It can be seen how easy it is to change the data and have the value immediately updated. Changing the number of parameters is a bit tricker, as it involves clearing the green region before changing the columns. It is not possible to change part of an array formula!

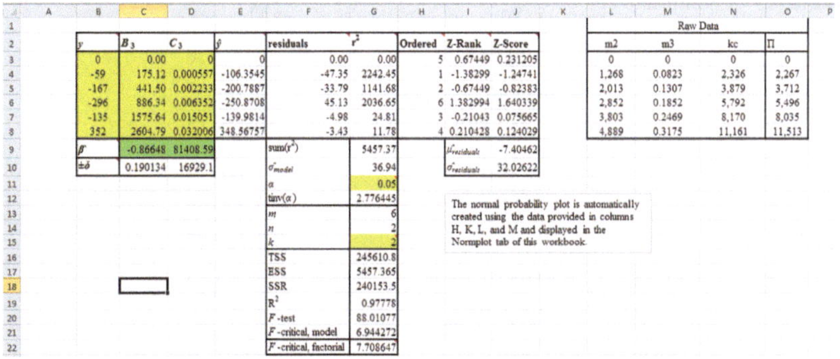

y	B_3	C_3	$\hat{y}$	residuals	r^2	Ordered	Z-Rank	Z-Score
0	0.00	0	0	0.00	0.00	5	0.67449	0.231205
-59	175.12	0.000557	-106.3545	-47.35	2242.45	1	-1.38299	-1.24741
-167	441.50	0.002233	-200.7887	-33.79	1141.68	2	-0.67449	-0.82383
-296	886.34	0.006352	-250.8708	45.13	2036.65	6	1.382994	1.640339
-135	1575.64	0.015051	-139.9814	-4.98	24.81	3	-0.21043	0.075665
352	2604.79	0.032006	348.56757	-3.43	11.78	4	0.210428	0.124029
β	-0.86648	81408.59		sum(r^2)	5457.37		$\mu_{residuals}$	-7.40462
$\pm\delta$	0.190134	16929.1		σ_{model}	36.94		$\sigma_{residuals}$	32.02622

α	0.05
tinv(α)	2.776445
m	6
n	2
k	2
TSS	245610.8
ESS	5457.365
SSR	240153.5
R^2	0.97778
F-test	88.01077
F-critical, model	6.944272
F-critical, factorial	7.708647

Raw Data

m2	m3	kc	Π
0	0	0	0
1,268	0.0823	2,326	2,267
2,013	0.1307	3,879	3,712
2,852	0.1852	5,792	5,496
3,803	0.2469	8,170	8,035
4,889	0.3175	11,161	11,513

The normal probability plot is automatically created using the data provided in columns H, K, L, and M and displayed in the Normplot tab of this workbook.

Fig. 4.6 Linear regression example: data analysis results after removing the outlier

4.2.2 Nonlinear Regression

Consider the problem of obtaining a relationship for the ratio between the equilibrium and isotonic cell volumes given the osmotic pressure. The theoretical relationship can be written as

$$\frac{V}{V_0} = (1 - b^*)\frac{-1 + \sqrt{1 + 4B\Pi_0}}{-1 + \sqrt{1 + 4B\Pi}} + b^* \tag{4.19}$$

where both B and b^* are the parameters to be determined and Π_0 is a known osmotic value. The experimental data is provided in Table 4.5. For this data set, Π_0 has a value of 0.293. Detailed information about the problem can be found in Ross-Rodriguez (2009). Data provided courtesy of Dr. Lisa Ross-Rodriguez.

Before we set up the problem in Excel, it is first necessary to compute some preliminary information. First, we need to obtain the derivatives of Eq. (4.19) with respect to the parameters, that is,

$$\frac{d(V/V_0)}{db^*} = 1 - \frac{1 - \sqrt{1 + 4B\Pi_0}}{1 - \sqrt{1 + 4B\Pi}} \tag{4.20}$$

$$\frac{d(V/V_0)}{dB} = 2(1 - b^*)\times \left[\frac{\Pi_0}{\sqrt{1 + 4B\Pi_0}\left(-1 + \sqrt{1 + 4B\Pi}\right)} - \frac{\Pi\left(-1 + \sqrt{1 + 4B\Pi_0}\right)}{\sqrt{1 + 4B\Pi}\left(-1 + \sqrt{1 + 4B\Pi}\right)^2}\right] \tag{4.21}$$

Table 4.5 Equilibrium cell volume data

V/Vo	Π
1.000 34	0.292 78
0.804 65	0.571 72
0.753 58	0.855 14
0.715 48	1.135 95
0.685 88	1.433 49
0.666 00	1.729 08
0.659 13	2.028 15
0.640 04	2.326 60
0.626 61	2.667 04

It can clearly be seen that this equation is nonlinear in the *parameters*. Thus, nonlinear regression using Solver will be performed. In order to obtain values for the parameter confidence intervals using Eq. (4.15), the grand Jacobian will be calculated using the "best" estimated values of the parameters and the above derivatives.

Next, open the nonlinear-regression template file nonlinearregression.xltm. Before we enter the data, it would be helpful to create the required macros so that we can immediately see the values (rather than some undefined error).

To create the macros, go to the Developer tab and click on the Visual Basic icon, as shown in Fig. 4.7. This will bring up a new window in which you can edit and create new visual basic code. We will need to update (or change) the functions dydb1, dydb3, and model. The updated versions are:

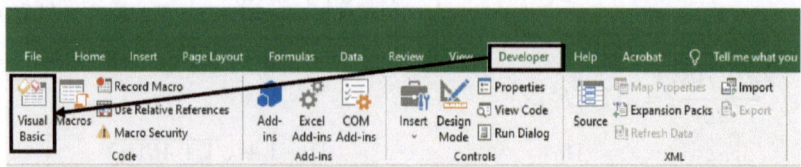

Fig. 4.7 Accessing the macros in Excel

```
Public Function model(parameter, x)
bs = parameter(1)
B = parameter(2)
model = (1 - bs) * (-1 + Sqr(1 + 4 * B * 0.293)) / (-1 + Sqr(1
+ 4 * B * x(1))) + bs
End Function
```

```
Function dydb1(parameter As Range, x As Range)
bs = parameter(1)
B = parameter(2)
dydb1 = 1 - (-1 + Sqr(1 + 4 * B * 0.293)) / (-1 + Sqr(1 + 4 *
B * x(1)))
End Function
```

```
Function dydb3(parameter As Range, x As Range)
bs = parameter(1)
B = parameter(2)
Pio = 0.293
so = Sqr(1 + 4 * B * Pio)
s = Sqr(1 + 4 * B * x(1))
dydb3 = 2 * (1 - bs) * (Pio / so / (s - 1) - x(1) * (so - 1) /
s / (s - 1) ^ 2)
End Function
```

Now that we have defined the functions, we can update the nonlinear regression template to match what we need. In our case, we will need to remove columns B, D and G as we only have two regressors, but a single source of data. Since we have 9 data points, but the template has space for 13 data points, we will delete rows 6, 7, 8, and 9. As a rule, one should always delete (and insert) rows and columns between the endpoints, for example, in our case, the A-range will change from being defined as F3:H15 to F3:H11 (after removing both the required columns and rows). This will allow for any ranges to be properly updated. Next, we will copy the output data (V/V_0) to the B column starting from B3. The data for Π will be copied into the C column starting from C3. If you want to check that everything has been correctly entered, then you will need to compare with some arbitrary parameter values. Let us select the values to be our initial guess, that is, 0.5 for b1 (b^*) and 2.5 for b2 (B). The results are shown in Fig. 4.8.

Once everything has been properly set up, go to the Data tab and select the Solver (see Fig. 4.6 for the details). A new window appears. It should be properly configured to solve the problem without making any changes. The default values are shown in Fig. 4.2. Once it has been confirmed that everything is correct, then press Solve. After a few seconds, another window will appear stating that a solution was found (hopefully, if everything has been correctly set up). Click OK and this solution will be saved. The final results are shown in Fig. 4.9.

Figure 4.10 shows the normal probability plot and a time-series plot of the residuals. It is easy to see that the B parameter is not significant and its value could be zero. Given the overall good fit and the relative well-behaved nature of the residuals, this would suggest that potentially not enough data have been collected to make an appropriate estimate. This situation partly explains why the Solver can have issues with obtaining a good value for B. The residual plots are

	B	C	D	E	F	G	H	I	J	K	L	M
				Jacobian								
y		x_3	$\partial y/\partial \beta_1$	$\partial y/\partial \beta_3$	$\hat{y}$	residuals	r^2		Ordered	Z-Rank	Z-Score		
	1.0003	0.29278	0.00	-0.00001	1.0003	0.00	0.00		8	0.967422	1.416873		
	0.8047	0.57172	0.38	0.00732	0.8086	0.00	0.00		9	1.593219	1.852022		
	0.7536	0.85514	0.53	0.0085	0.7350	-0.02	0.00		3	-0.58946	-0.59921		
	0.7155	1.13595	0.61	0.008591	0.6953	-0.02	0.00		2	-0.96742	-0.77349		
	0.6859	1.43349	0.66	0.008391	0.6685	-0.02	0.00		4	-0.28222	-0.47037		
	0.6660	1.72908	0.70	0.008113	0.6499	-0.02	0.00		5	0	-0.32563		
	0.6591	2.02815	0.73	0.007822	0.6359	-0.02	0.00		1	-1.59322	-1.09577		
	0.6400	2.3266	0.75	0.007543	0.6251	-0.01	0.00		6	0.282216	-0.19862		
	0.6266	2.66704	0.77	0.007248	0.6153	-0.01	0.00		7	0.589456	0.194185		
		β^*	0.5	2.5		sum(r^2)		0.00		$\mu_{residuals}$	-0.0131		
		$\pm\delta$	0.11255	9.239356		σ_{model}		0.02		$\sigma_{residuals}$	0.009203		
						a		0.05					
						tinv(a)		2.364624		The normal probability plot is automatically			
						m		9		created using the data provided in columns			
						n		2		H, L, M, and N and displayed in the			
						k		2		Normplot tab of this workbook.			
						TSS		0.109233					
						SSE		0.002221					
						SSR		0.107012					
						R^2		0.979666					
						F-test		168.6259					
						F-critical, model		4.737414					
						F-critical, factorial		5.591448					

Fig. 4.8 Problem set-up using initial parameter values

	V/Vo	Π	$\partial y/\partial \beta_1$	$\partial y/\partial \beta_2$	ŷ	residuals	r^2	Ordered	Z-Rank	Z-Score
			Jacobian							
	1.0003	0.29278	0.00	-0.00001	1.0003	0.00	0.00	5	0	-0.05602
	0.8047	0.57172	0.38	0.007262	0.8174	0.01	0.00	9	1.593219	1.924833
	0.7536	0.85514	0.53	0.008435	0.7472	-0.01	0.00	1	-1.59322	-1.02829
	0.7155	1.13595	0.61	0.008527	0.7095	-0.01	0.00	2	-0.96742	-0.9756
	0.6859	1.43349	0.66	0.008329	0.6840	0.00	0.00	4	-0.28222	-0.33586
	0.6660	1.72908	0.70	0.008053	0.6664	0.00	0.00	6	0.282216	0.015948
	0.6591	2.02815	0.73	0.007763	0.6531	-0.01	0.00	3	-0.58946	-0.97267
	0.6400	2.3266	0.75	0.007486	0.6429	0.00	0.00	7	0.589456	0.3941
	0.6266	2.66704	0.77	0.007194	0.6336	0.01	0.00	8	0.967422	1.03356
		β	0.524581	2.408129		sum(r^2)	0.00	$\mu_{residuals}$		0.000291
		$\pm\delta$	0.043624	3.616267		σ_{model}	0.01	$\sigma_{residuals}$		0.006449
						α	0.05			
						tinv(α)	2.364624			
						m	9			
						n	2			
						k	2			
						TSS	0.109233			
						ESS	0.000333			
						SSR	0.1089			
						R^2	0.996947			
						F-test	1143.092			
						F-critical, model	4.737414			
						F-critical, factorial	5.591448			

The normal probability plot is automatically created using the data provided in columns H, K, L, and M and displayed in the Normplot tab of this workbook.

Fig. 4.9 Nonlinear regression example: Excel spreadsheet results

shown in Fig. 4.10. Overall, the results are decent, given the small sample. Since it has been assumed that the given equation holds, in order to obtain a better understanding of the data, additional experiments should be performed.

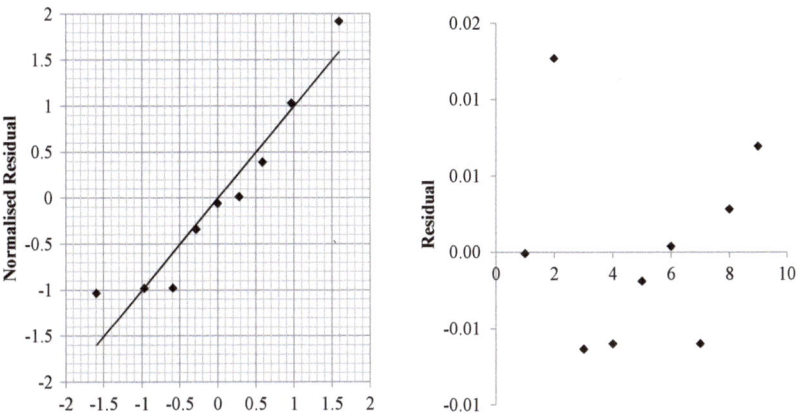

Fig. 4.10 Nonlinear regression example: (left) normal probability plot and (right) time-series plot of the residuals

4.2.3   Factorial Design Example

Consider the problem of optimising a distillation column to determine the effects of different parameters on the overall purity of the overhead product. The variables of interest are reboiler duty (A), feed temperature (B), reflux ratio (C), and feed location (D). The purity of the product is expressed in a proprietary scale where 150 is absolutely pure and 50 is 70% pure. The data obtained from this 2^4-factorial experiment with no replicates are shown in Table 4.6. Analyse the data to determine which parameters are significant and what the final model could be. The full model of interest can be written as

$$
\begin{aligned}
y = {} & \beta_0 + \beta_1 x_1 + \beta_2 x_2 + \beta_3 x_3 + \beta_4 x_4 + \beta_{12} x_1 x_2 + \beta_{13} x_1 x_2 + \beta_{14} x_1 x_4 \\
& + \beta_{23} x_2 x_3 + \beta_{24} x_2 x_4 + \beta_{34} x_3 x_4 + \beta_{123} x_1 x_2 x_3 + \beta_{124} x_1 x_2 x_4 \\
& + \beta_{134} x_1 x_3 x_4 + \beta_{234} x_2 x_3 x_4 + \beta_{1234} x_1 x_2 x_3 x_4
\end{aligned}
\tag{4.22}
$$

Table 4.6 Factorial design data for a plant distillation column

y	A (x_1)	B (x_2)	C (x_3)	D (x_4)
45	−1	−1	−1	−1
71	1	−1	−1	−1
48	−1	1	−1	−1
65	1	1	−1	−1
68	−1	−1	1	−1
60	1	−1	1	−1
80	−1	1	1	−1
65	1	1	1	−1
43	−1	−1	−1	1
100	1	−1	−1	1
45	−1	1	−1	1
104	1	1	−1	1
75	−1	−1	1	1
86	1	−1	1	1
70	−1	1	1	1
96	1	1	1	1

Additional details and information about this question can be found in Example 4.2 in the book.

In order to solve this problem, we will use the Excel template (factorialdesigntemplate.xltm). After opening the file, the first step is to make sure that the macros have been enabled. Once that has been completed, then we need to make sure that we have the required numbers of columns and rows in the yellow region. For a 2^4-full factorial design, we will have 16 parameters. From Table 4.6, we see that we also have 16 data points. Since the default template has 8 parameters and 16 data points, we will need to add 8 extra columns to the default template. The easiest way to do this is to first select 4 columns (say E, F, G, and H; never selecting the first or last column), and right-clicking. In the menu that appears, select Insert and Excel will immediately insert 4 more columns.[5] Repeat once more. We will also need to add the missing formulae at the bottom of the spreadsheet. Select cell D20 and drag it to M20. Select cells D23:D28 and drag them across to column M.

If desired, we can update the names of the parameters to match that of the model. This is helpful if we wish to keep track of which parameter is which. Following the same order as in the model given above is the simplest solution.

To enter the data, the simplest solution is to copy it from Table 4.6. The higher-order interactions can be obtained by multiplying the relevant columns of the data. In column B starting at B3, copy and paste the y data. In columns D, E, F, and G, place the x_1, x_2, x_3, and x_4 data (always starting from the third row). Note that column C should be a column of 16 ones! For the remaining columns, enter the following formulae in the third row:

[H3]	=D3*E3	=1
[I3]	=D3*F3	=1
[J3]	=D3*G3	=1
[K3]	=E3*F3	=1
[L3]	=E3*G3	=1
[M3]	=F3*G3	=1
[N3]	=D3*E3*F3	=-1
[O3]	=D3*E3*G3	=-1
[P3]	=D3*F3*G3	=-1
[Q3]	=E3*F3*G3	=-1
[R3]	=D3*E3*F3*G3	=1

---

[5] In general, if you select n columns or rows, then Excel will insert n more columns or rows.

Once, all the formulae have been entered into the third row, select cells H3:R3 and drag it down to row 18. You will probably want to select the Fill without Formatting option when dragging down the values. All the formulae should then update based on the relevant values. The last step is to select the green region C19:R19 starting from C19. Once done, click in the formula window, and then, press CTRL + SHIFT + ENTER and the array formula in C19 will be copied to all the remaining cells. The final result is shown in Fig. 4.11.

It should be noted that since we have 16 parameters and 16 data points, we cannot estimate the variances so anything requiring a variance will a #NUM! or #DIV/0! error message. However, we can still determine which parameters are significant by using a normal probability plot of the parameters. The template creates this plot automatically and places it in the tab Normplot Parameters. Switching to this tab, should show a figure similar to that in Fig. 4.12. Visually, since we seek those parameters that are *not* normally distributed, the points lying between an x-value of -1.5 and 0.5 are normal, and hence, can be excluded as significant parameters. The easiest way to determine which parameters are significant is to record the ordinal value of the significant points starting from the smallest (most negative) value. In our case, it appears that the parameters corresponding to rank 1, 12, 13, 14, 15, and 16 are significant. To determine which these parameters are, we can return to the Regression tab and look at row 23, where we have the label Ordered, which ranks our parameters from smallest to largest. Therefore, the parameter with rank 1 corresponds to β_{13}; to rank 12, β_3; to rank 13, β_4; to rank 14, β_{14}; to rank 15, β_1; and to rank 16, β_0. Thus, the simplified model can be written as

$$y = 70.1 + 10.8x_1 + 4.94x_3 + 7.31x_4 - 9.06x_1x_3 + 8.31x_1x_4 \qquad (4.23)$$

In order to analyse this reduced model, we will need to remove the irrelevant columns. Unfortunately, since we have used an array formula in the green region, we cannot just simply delete the columns. First, we need to remove the array formula. Select the region C19:R19. Copy the array formula in the formula window. Once the formula has been copied, press DEL to clear the array formula. Select C19 and paste the copied formula. Press CTRL + SHIFT + ENTER. This allows us to save the formula for future use.

Select columns E, H, K, L, M, N, O, P, Q, and R. Right-click and select Delete. We now need to restore the parameter values in the green region. The procedure is the same as was done initially. Select the region C19:H19, click in the formula window, and press CTRL + SHIFT + ENTER. The result is shown in Fig. 4.13. Note that the parameter values obtained should be identical to those

y	β_0	β_1	β_3	β_4	β_{12}	β_{13}	β_{14}	β_{23}	β_{24}	β_{34}	β_{123}	β_{124}	β_{134}	β_{234}	β_{1234}	$\hat{y}$
45																
71																
48																
65																
68																
60																
80																
65																
43																
100																
45																
104																
75																
86																
70																
96																
β	70.0625	10.8125	1.5625	4.9375	7.3125	0.0625	-9.0625	8.3125	1.1875	-0.1875	-0.5625	0.9375	2.0625	-0.8125	-1.3125	0.6875
$t \cdot \beta$	#NUM!	#NUM!	#NUM!	#NUM!	#NUM!	#NUM!	#NUM!	#NUM!	#NUM!	#NUM!	#NUM!	#NUM!	#NUM!	#NUM!	#NUM!	#NUM!

Factorial Analysis of Parameters

Ordered	16	15	10	12	13	6	1	14	9	5	4	8	11	3	2	7
Z-Rank	1.862732	1.318011	0.237202	0.579132	0.776422	-0.40225	-1.86273	1.00999	0.078412	-0.57913	-0.77642	-0.28627	0.40225	-1.00999	-1.31801	-0.2372
Z-Score	3.622594	0.272136	-0.25093	-0.06008	0.074219	-0.33575	-0.85175	0.130767	0.078412	-0.57913	-0.34989	-0.3711	-0.22266	-0.38523	-0.41351	-0.30041
SSR_i	#NUM!	#NUM!	#NUM!	#NUM!	#NUM!	#NUM!	#NUM!	#NUM!	#NUM!	#NUM!	#NUM!	#NUM!	#NUM!	#NUM!	#NUM!	#NUM!
F_i	#NUM!	#NUM!	#NUM!	#NUM!	#NUM!	#NUM!	#NUM!	#NUM!	#NUM!	#NUM!	#NUM!	#NUM!	#NUM!	#NUM!	#NUM!	#NUM!
Decision	#NUM!	#NUM!	#NUM!	#NUM!	#NUM!	#NUM!	#NUM!	#NUM!	#NUM!	#NUM!	#NUM!	#NUM!	#NUM!	#NUM!	#NUM!	#NUM!

$n_{parameters}$	6
$\sigma_{parameters}$	17.68415

	residuals	r^2		Ordered	Z-Rank	Z-Score
45	0.00	0.00		1	-1.86273	#DIV/0!
71	0.00	0.00		1	-1.86273	#DIV/0!
48	0.00	0.00		1	-1.86273	#DIV/0!
65	0.00	0.00		1	-1.86273	#DIV/0!
68	0.00	0.00		1	-1.86273	#DIV/0!
60	0.00	0.00		1	-1.86273	#DIV/0!
80	0.00	0.00		1	-1.86273	#DIV/0!
65	0.00	0.00		1	-1.86273	#DIV/0!
43	0.00	0.00		1	-1.86273	#DIV/0!
100	0.00	0.00		1	-1.86273	#DIV/0!
45	0.00	0.00		1	-1.86273	#DIV/0!
104	0.00	0.00		1	-1.86273	#DIV/0!
75	0.00	0.00		1	-1.86273	#DIV/0!
86	0.00	0.00		1	-1.86273	#DIV/0!
70	0.00	0.00		1	-1.86273	#DIV/0!
96	0.00	0.00		1	-1.86273	#DIV/0!
sum(r^2)	0.00	#DIV/0!		$\mu_{residuals}$		0
α		0.05		$\sigma_{residuals}$		0
lin(α)		#NUM!				
m		16				
n_R		2				
k		15				
TSS		5730.938				
SSE		#DIV/0!				
SSR		#DIV/0!				
R^2		#DIV/0!				
F-test		#NUM!				
F-critical model		#NUM!				
F-critical factorial		#NUM!				

Fig. 4.11 Factorial design in Excel

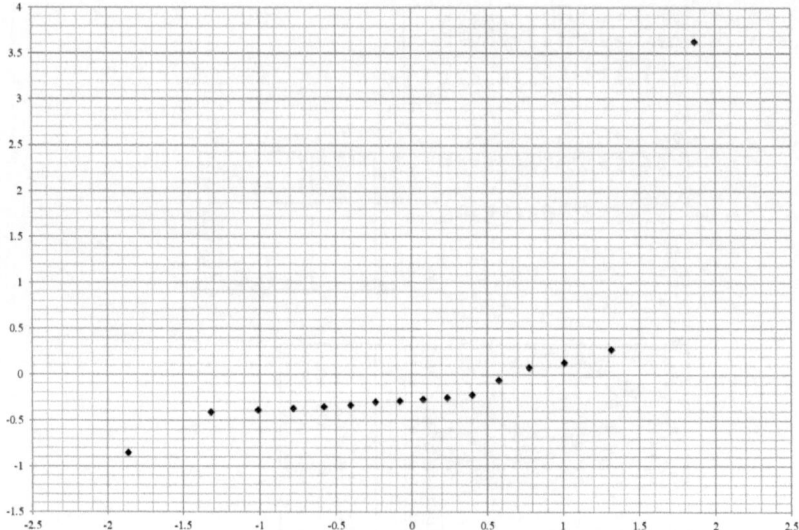

Fig. 4.12 Normal probability plot for factorial design

previous obtained. We can then examine the different tabs to see if the model is sufficient. The normal probability plot and the residuals *vs.* y are shown in Fig. 4.14. From Fig. 4.13, we can see that R^2 is 96.6%, which suggests that the model captures much of the variance present. Figure 4.15 shows that the residuals are normally distributed as there are no obvious deviations in the normal probability plot. The plot of the residuals as a function of the original data seems to suggest that for data points between 60 and 100 have much greater variability than those less than 60. Given the lack of replicates, there is little more that can be said.

Let us now return to the full model and try something different. We can note that our reduced model does not contain any parameters with factor x_2. This suggests that this factor is not significant and does not influence the results. Since a complete factor can be removed, we can use the property of projection, which basically states that removing a factor leaves us with a factorial design with replicates, to analyse the results from a different perspective. Therefore, let us return to the full-model Excel spreadsheet (pressing CTRL + Z thrice will do the trick if you have not made any interim changes to the above procedure). Now, we will remove only those columns corresponding to the factor x_2.

y	β_0	β_1	β_3	β_4	β_{13}	β_{14}	$\hat{y}$	residuals	r^2	Ordered	Z-Rank	Z-Score
45	1	-1	-1	-1	1	1	46.25	1.25	1.56	11	0.40225	0.346577
71	1	1	-1	-1	-1	-1	69.375	-1.63	2.64	7	-0.2372	-0.45055
48	1	-1	-1	-1	1	1	46.25	-1.75	3.06	6	-0.40225	-0.48521
65	1	1	-1	-1	-1	-1	69.375	4.38	19.14	14	1.00999	1.213018
68	1	-1	1	-1	-1	1	74.25	6.25	39.06	15	1.318011	1.732883
60	1	1	1	-1	1	-1	61.125	1.13	1.27	10	0.237202	0.311919
80	1	-1	1	-1	-1	1	74.25	-5.75	33.06	1	-1.86273	-1.59425
65	1	1	1	-1	1	-1	61.125	-3.88	15.02	2	-1.31801	-1.07439
43	1	-1	-1	1	1	-1	44.25	1.25	1.56	11	0.40225	0.346577
100	1	1	-1	1	-1	1	100.625	0.63	0.39	9	0.078412	0.173288
45	1	-1	-1	1	1	-1	44.25	-0.75	0.56	8	-0.07841	-0.20795
104	1	1	-1	1	-1	1	100.625	-3.38	11.39	4	-0.77642	-0.93576
75	1	-1	1	1	-1	-1	72.25	-2.75	7.56	5	-0.57913	-0.76247
86	1	1	1	1	1	1	92.375	6.38	40.64	16	1.862732	1.76754
70	1	-1	1	1	-1	-1	72.25	2.25	5.06	13	0.776422	0.623838
96	1	1	1	1	1	1	92.375	-3.63	13.14	3	-1.00999	-1.00507

β	70.0625	10.8125	4.9375	7.3125	-9.0625	8.3125	sum(r^2)	195.13	$\mu_{residuals}$	0
$\pm\delta$	2.460587	2.460587	2.460587	2.460587	2.460587	2.460587	σ_{model}	4.42	$\sigma_{residuals}$	3.606707
							α	0.05		

Factorial Analysis of Parameters								tinv(α)	2.228139
Ordered	6	5	2	3	1	4		m	16
Z-Rank	1.382994	0.67449	-0.67449	-0.21043	-1.38299	0.210428		n	6
Z-Score	1.974529	-0.16555	-0.37775	-0.29197	-0.88342	-0.25585		n_R	2
SSR_i	78540.06	1870.563	390.0625	855.5625	1314.063	1105.563		k	5
F_i	4025.115	95.86483	19.99039	43.84689	67.34465	56.65919		TSS	5730.938
Decision	Keep	Keep	Keep	Keep	Keep	Keep		SSE	195.125
								SSR	5535.813
$\mu_{parameters}$	15.39583							R^2	0.965952
$\sigma_{parameters}$	27.68592							F-test	56.74119
								F-critical, model	3.325835
								F-critical, factorial	4.964603

Fig. 4.13   Analysis of the reduced factorial model

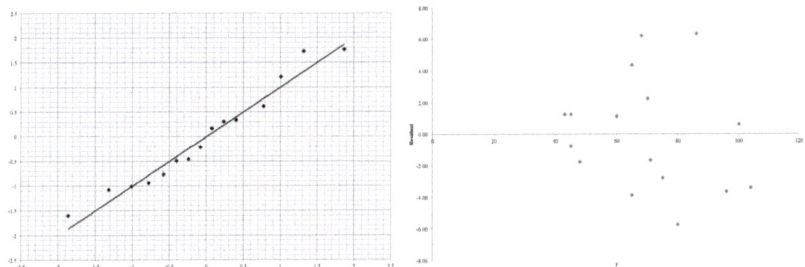

Fig. 4.14   (Left) Normal probability plot and (right) residuals as a function of y for the reduced factorial model

However, before proceeding, we need to make sure that there are no array formulae. Select the region C19:R19. Copy the array formula in the formula window. Once the formula has been copied, press DEL to clear the array formula.

y	β_0	β_1	β_2	β_4	β_{12}	β_{14}	β_{34}	β_{134}	$\hat{y}$	residuals	r^2	Ordered	Z-Rank	Z-Score
45	1	-1	-1	-1	1	1	1	-1	46.5	1.50	2.25	10	0.237202	0.433615
71	1	1	-1	-1	-1	-1	1	1	68	-3.00	9.00	3	-1.00999	-0.86723
48	1	-1	-1	-1	1	1	1	-1	46.5	-1.50	2.25	7	-0.2372	-0.43362
65	1	1	-1	-1	-1	-1	1	1	68	3.00	9.00	14	1.00999	0.867231
68	1	-1	1	-1	-1	1	-1	1	74	6.00	36.00	16	1.862732	1.734461
60	1	1	1	-1	1	-1	-1	-1	62.5	2.50	6.25	12	0.579132	0.722692
80	1	-1	1	-1	-1	1	-1	1	74	-6.00	36.00	1	-1.86273	-1.73446
65	1	1	1	-1	1	-1	-1	-1	62.5	-2.50	6.25	4	-0.77642	-0.72269
43	1	-1	-1	1	1	-1	-1	1	44	1.00	1.00	9	0.078412	0.289077
100	1	1	-1	1	-1	1	-1	-1	102	2.00	4.00	11	0.40225	0.578154
45	1	-1	-1	1	1	-1	-1	1	44	-1.00	1.00	8	-0.07841	-0.28908
104	1	1	-1	1	-1	1	-1	-1	102	-2.00	4.00	6	-0.40225	-0.57815
75	1	-1	1	1	-1	-1	1	-1	72.5	-2.50	6.25	4	-0.77642	-0.72269
86	1	1	1	1	1	1	1	1	91	5.00	25.00	15	1.318011	1.445385
70	1	-1	1	1	-1	-1	1	-1	72.5	2.50	6.25	12	0.579132	0.722692
96	1	1	1	1	1	1	1	1	91	-5.00	25.00	2	-1.31801	-1.44538

β	70.0625	10.8125	4.9375	7.3125	-9.0625	8.3125	-0.5625	-0.8125	sum(r^2)	179.50	$\mu_{residuals}$	0
$\pm\delta$	2.730784	2.730784	2.730784	2.730784	2.730784	2.730784	2.730784	2.730784	σ_{model}	4.74	$\sigma_{residuals}$	3.459287

									α	0.05
									tinv(α)	2.306004

Factorial Analysis of Parameters										
Ordered	8	7	4	5	1	6	3	2	m	16
Z-Rank	1.534121	0.887147	-0.15731	0.157311	-1.53412	0.488776	-0.48878	-0.88715	n	8
Z-Score	2.390055	-0.02291	-0.26217	-0.16545	-0.83232	-0.12472	-0.48616	-0.49634	n_R	2
SSR_i	78540.06	1870.563	390.0625	855.5625	1314.063	1105.563	5.0625	10.5625	k	7
F_i	3500.393	83.36769	17.3844	38.13092	58.56546	49.27298	0.225627	0.470752	TSS	5730.938
Decision	Keep	Keep	Keep	Keep	Keep	Keep	Discard	Discard	SSE	179.5

$\mu_{parameters}$	11.375	SSR	5551.438
$\sigma_{parameters}$	24.55488	R^2	0.968679
		F-test	35.3454
		F-critical, model	3.500464
		F-critical, factorial	5.317655

Fig. 4.15   Result after removing factor x_2

Select C19 and paste the copied formula. Press CTRL + SHIFT + ENTER. This allows us to save the formula for future use.

Next, select columns E, H, K, L, N, O, Q, and R, which all contain x_2. Right-click and select Delete. We now need to restore the parameter values in the green region. The procedure is the same as was done initially. Select the region C19:J19, click in the formula window, and press CTRL + SHIFT + ENTER. The result is shown in Fig. 4.15. Note that the parameter values obtained should be identical to those previous obtained. We now have a 2^3-full factorial design with two replicates ($n_R = 2$). This implies that we can now compute the variance for our model and analyse the significant parameters using a different approach, namely using the SSR_i. This analysis is shown in the area labelled Factorial Analysis of Parameters, specifically rows 26, 27, and 28. In row 28, the computer automatically gives the decision based on a comparison between F_i and the critical value of the F-test for the factorial model (given in cell M33). We see once again that the parameters labelled KEEP are the same as before. Therefore, this approach provides us with essentially the same results as before and shows that we have found the best model given the data.

It should be noted that in general the different approach to determining significant parameters need not provide exactly the same results. However, they should be similar.

4.3 Practice Questions

Solve the following questions using Excel.

(1) Consider fitting the Antoine Equation to some vapour pressure as a function of temperature data that was obtained using toluene given in Table 4.7. The general form of the Antoine Equation can be written as

$$P^{vap} = 10^{A + \frac{B}{C+T}} \qquad (4.24)$$

where A, B, and C are parameters, T is the temperature in °C, and P^{vap} is the vapour pressure of toluene in mm Hg. Two separate runs were performed using two different makes of measurement devices. By fitting a linearised model, a nonlinear model obtained by taking the $\log_{10}$ of Eq. (4.24), and a nonlinear model obtained using Eq. (4.24) to the data, and analysing the residuals, answer the following questions:

Table 4.7 Partial pressures of toluene at different temperatures (for Question 1)	Temperature, T (°C)	Vapour Pressure, P^{vap} (mm Hg)	
		Run 1	Run 2
	−4.4	5.05	5.15
	6.4	10.0	9.89
	18.4	20.1	21.9
	31.8	39.9	40.8
	40.3	59.8	62.5
	51.9	99.9	97.8
	69.5	200	206
	89.5	400	415
	110.6	760	747
	136.5	1502	1512

a. Are the errors for the two runs the same? How can this be determined?
b. Obtain separate parameter estimates for each of the runs and models. Which model best describes the data for the given run? What does this suggest about the appropriate error structure for each run?
c. Using the best parameter estimates for A, B, and C, compare them against the theoretical values of $A = 6.954\ 64$, $B = -1,344.8\ °C$, and $C = 219.482\ °C$ (Dean 1999). Are the experimental values close to the accepted values?

   Hint: For the nonlinear models, it is suggested that the estimates obtained using the linearised model be used as the initial guess for the nonlinear method.

(B) Consider the problem of trying to determine which conditions impact the life (in hours) of a machine tool. The variables of interested have been selected as: cutting speed (A), tool geometry (B), and cutting angle (C). Consider the following full factorial design whose regression matrix and results are shown in Table 4.8. Perform all analysis at the 95% level. Answer the following questions:
   (a) Determine the model for the full factorial experiment.
   (b) Fit the model and obtain confidence intervals for the parameter estimates. Determine which parameter estimates should be kept.
   (c) Calculate the F-score for each parameter estimate. Determine which parameters should be kept.
   (d) Are the results from (b) and (c) the same? Do you think that this is a coincidence or will this always be the case?
   (e) Based on your results from (b) and (c) what model would you suggest for the data? Which interactions are significant? Why?
   (f) Examine the residuals for the full model and determine if there are any issues with the distribution of the residuals. (*Hint: Plot the residuals for each replicate in a different colour or on separate graphs.*)

Table 4.8 Tool life data (for Question 2)

Run	A	B	C	Replicate		
				y_1	y_2	y_3
1	−	−	−	22	31	25
2	+	−	−	32	43	29
3	−	+	−	35	34	50
4	+	+	−	55	47	46
5	−	−	+	44	45	38
6	+	−	+	40	37	36
7	−	+	+	60	50	54
8	+	+	+	39	41	47

Data taken from D. Montgomery (2007), *Design and Analysis of Experiments*, 6th Edition (2007)

References

Dean, J. A. (1999). *Lange's Handbook of Chemistry* (15th ed.). New York, New York, United States of America: McGraw-Hill, Inc.

Prickett, R. C., Elliott, J. A., & McGann, L. E. (2011). Application of the Multisolute Osmotic Virial Equation to Solutions Containing Electrolytes. *The Journal of Physical Chemistry B, 115*, 14531–14543.

Ross-Rodriguez, L. U. (2009). Cellular Osmotic Properties and Cellular Responses to Cooling. Edmonton, Alberta, Canada: University of Alberta.

References